Generis
PUBLISHING

Geosciences With New Eyes:
Not Just *Stones And Disasters*

Susanna Occhipinti

Title: Geosciences With New Eyes:
Not Just *Stones And Disasters*

ISBN: 979-8-89966-330-7

Author: Susanna Occhipinti

Cover image: Susanna Occhipinti

Publisher: Generis Publishing
Online orders: www.generis-publishing.com
Contact email: info@generis-publishing.com

GEOSCIENCES WITH NEW EYES: NOT JUST *STONES* AND *DISASTERS*

Reflections and methodologies to help students understand Earth processes, stimulate curiosity, and educate citizens who are aware of the environmental challenges of the future

Finding an adequate title for this book was not easy. I did not want to appear presumptuous in believing that this work could "revolutionize" geoscience teaching; however, I felt the desire to share my ideas and experiences gained over the years. I thought of borrowing from Proust the idea of "looking at the world with ever-new eyes," because I would like this book to represent a journey into geosciences with new eyes. The terms "rocks and catastrophes," often used in my research, seemed natural for the title.

This text is the result of a reworking of numerous articles, to which I have tried to give coherence and continuity, with the hope that it may be useful to those who believe in the educational potential of geosciences. Nearly all the educational and methodological suggestions presented have been field-tested on numerous occasions. The outcomes were generally positive, though critical observations allowed me to refine the proposals.

I hope that my profound passion for this discipline emerges more than anything else—a discipline so neglected in schools yet fundamental for understanding the world and facing the challenges of contemporary society. Geosciences are not merely "rocks and catastrophes," but constitute a key for interpreting the past, understanding the present, and imagining a sustainable future for our planet.

I thank all those who have helped me grow professionally and personally, who have allowed me to experiment in their classrooms and have recognised my commitment. I do not list you one by one because you are numerous but know that this work would not have been possible without you.

If even just one teacher, reading these pages, finds inspiration to renew their educational approach, or if one student discovers the beauty of geosciences, my objective will be achieved. Educating in geosciences means forming aware citizens, capable of understanding the complexity of the Earth system.

Susanna Occhipinti

Susanna Occhipinti's book is a valuable tool for teachers, educators, and all geoscience enthusiasts. It summarizes the decades of experience of a passionate and extremely competent geology teacher. Works of this kind are still rare in Italy, despite the growing need for innovative teaching resources.

Geology is a fascinating but complex discipline that integrates basic skills in mathematics, physics, chemistry, computer science, and other sciences to study and interpret the natural phenomena of our planet and the solar system. As the author emphasizes from the very first pages, geology is not limited to the study of 'rocks' or natural disasters—a reductive and uninspiring image that is often conveyed to young people—but is a rich and complex science.

Through this book, the author accompanies the reader on a journey of discovery of the variety and depth of geology, illustrating often unknown but highly interesting aspects that can inspire a fresher and more engaging approach to teaching. However, it is not enough to know the content of geology: the book also offers practical ideas, suggestions, and concrete tools for teaching this fascinating subject in a more modern and compelling way.

Teaching geology, understanding how the Earth works today and how it worked in the past, is essential for interpreting our future and educating citizens who are more aware of the territory and its geodiversity, which also includes multiple types of risk. The answer to these challenges is knowledge, and this book is a valuable guide along this path.

Enjoy the read!

Rodolfo Carosi

President of the Italian Geological Society

Notes on images, quotations, and resources

At the end of each chapter, there is a photograph taken by the author during her travels and geological explorations. These images are not merely decorative elements, but visual testimonies of a personal journey through the landscapes and phenomena discussed in the preceding pages.

The decision to include these photographs stems from the belief that Geosciences are not just an academic discipline, but also an aesthetic and emotional experience. Geological landscapes hold a special beauty that comes from an awareness of the processes that created them and the possibility of 'reading' the history of our planet in them.

These images therefore invite the reader not to limit themselves to the theoretical dimension of the study, but to explore the territory personally, developing that conscious gaze that transforms a simple landscape into an open book on the history of the Earth.

The essential captions accompanying the images are intended simply to locate the subject, leaving the reader free to establish connections with the contents of the chapter and to appreciate the intrinsic beauty of the geological phenomena represented.

The other images are taken from Wikipedia or Creative Commons and are therefore free to use.
I apologize for any errors or omissions in case I have forgotten to cite the authors.

All citations contained in the text are listed in the references at the end of each specific chapter, in the "Further reading" box.

At the end of the book, there are USEFUL LINKS to WEBSITES, RESOURCES, and BIBLIOGRAPHY

This book draws upon research and teaching experience from multiple educational contexts, though specific examples and case studies naturally reflect the author's direct experience. The principles, methodologies, and challenges discussed apply broadly across diverse educational systems, and readers are encouraged to adapt approaches to their specific geographical, cultural, and institutional contexts. Where examples reference specific locations or systems, they serve to illustrate universal principles applicable globally.

Table of Contents

Introduction

This book brings together contributions developed over years of teaching, research, and experimentation in numerous classrooms, seeking to address the challenges of teaching geosciences from different perspectives and with innovative tools. Without claiming to be revolutionary, it offers alternative approaches to make geosciences more engaging and meaningful in the contemporary educational landscape, integrating personal reflections and in-depth research: the aim is to renew the perception of this discipline, which is often relegated to the margins of the school curriculum.

The following pages offer a journey through teaching methodologies and specific and cross-curricular skills, venturing into epistemology and exploring unexpected links between geosciences and other disciplines, often revealing surprising connections: geological phenomena that have influenced human history, planetary processes intertwined with social dynamics, seemingly ordinary minerals at the root of technological revolutions.

The aim is to provide teachers, educators, and enthusiasts with tools to transform the teaching of geosciences from the simple transmission of knowledge to an experience of discovery and understanding of the Earth system in all its fascinating complexity.

CHAPTER 1.
RETHINKING THE TEACHING OF GEOSCIENCES: FROM DISCIPLINARY FRAGMENTATION TO INTEGRATED TEACHING

A discipline in search of identity

Earth sciences, or geosciences, are currently in a unique position in the scientific education landscape. Their relevance has never been more evident: climate change, natural hazards, resource management, and environmental sustainability are central issues in public debate and international political agendas. On the other hand, this discipline continues to occupy a marginal position in school curricula, often relegated to a secondary role compared to other scientific disciplines, despite its educational potential and social relevance.

While the labour market is seeing growing demand for geoscience skills—from environmental consulting to critical mineral exploration, natural hazard assessment to renewable energy— enrolment in geoscience university courses is declining steadily and significantly. This phenomenon, which is particularly worrying for traditional geology courses, risks creating a skills gap at a time when society needs them most.

This paradox reflects broader difficulty recognising and communicating the true nature of geosciences: not simply a collection of facts about rocks and natural phenomena, but a fundamental key to understanding our planet as a complex and dynamic system. The traditional perception of this science fails to highlight its methodological complexity, its practical value, and its technologically advanced nature, which make it essential to contemporary science and economics, especially in emerging sectors. These aspects will be explored in Chapter 2, 'Geology, Earth Sciences, or Geosciences: Let's Try to Clarify,' where we will address the issues of terminology and the disciplinary identity of modern geosciences.

1.1 Dynamics of interest: when other disciplines prevail

In the landscape of scientific disciplines, both in school curricula and in university orientation, geosciences often start at a disadvantage. In the competition between STEM subjects, they are regularly surpassed by medicine, computer science, physics, and mathematics, but also by biology and biotechnology. The latter have managed to build a more modern and innovative image: they promise revolutionary discoveries and evoke the image of the researcher in a white coat with a micropipette in their hand.

It is a common experience to see normally attentive classes yawning in front of trays of rock samples, or brilliant students in national Earth Science competitions completely lost in front of the incomprehensible 'stains' on a geological map. Even more significant is the phenomenon of students who win prizes in international competitions but, when it comes to choosing a university, opt for faculties considered scientifically 'more serious'. For them, geosciences remain a second-tier option, unable to offer stimulating or economically competitive career prospects.

There are several reasons for this. Medical and biotechnological innovations attract more attention because they have an immediate impact on people's lives. Infectious diseases are perceived as immediate and universal threats: just think of how the COVID-19 pandemic has affected the collective imagination much more deeply than geological disasters that have caused a similar number of victims.

The media also play a decisive role. Geological events, however devastating, affect specific areas and receive intense but temporary attention, quickly disappearing from the spotlight once the emergency is over. The NIMBY (Not In My Backyard) principle applies: if it does not directly affect our territory, interest drops dramatically. A health threat, on the other hand, keeps attention high everywhere.

Biological research offers tangible and immediate results: new therapies, innovative drugs, agricultural practices that improve daily life. The biotechnology and pharmaceutical sectors receive large investments, promote rapid progress, and know how to communicate it effectively. Even in the climate change debate, biological aspects receive greater emphasis because of their direct link to human health.

Communication strategies contribute significantly to this imbalance: biologists have developed effective ways to make complex topics accessible and engaging. The intrinsic complexity of geosciences, on the other hand, is rarely communicated with equal effectiveness, thus remaining distant from everyday experience.

Finally, perceptions about job opportunities also play a role. Students choose paths that seem to be the most promising professionally, ignoring the growing demand for geoscience experts in innovative and strategic sectors.

This combination of factors represents a major challenge for geoscience education and requires innovative teaching approaches, and a comprehensive rethinking of how the discipline is presented.

1.2 Teaching geosciences: curriculum and reality

The teaching of geosciences in schools worldwide presents specific critical issues conditioning its effectiveness and perception. Despite geological content being widely present in national curricula across many countries, the discipline faces similar global challenges globally. The first concerns teacher training: in most educational systems, geology teachers often come from predominantly non-geological backgrounds, whether in biology, chemistry, or general science. Without a solid geological foundation, it is natural for educators to feel less confident about certain topics and to favour the more descriptive aspects of the subject. This situation represents a widespread pattern across educational systems worldwide.

A second challenge relates to the very nature of geological processes, which unfold over times and spaces difficult to reproduce in traditional educational contexts. Unlike biology or chemistry, where many phenomena can be observed and experimented with directly in the laboratory, geosciences require specific educational approaches. This peculiarity, apparently limiting, can instead become an opportunity to develop innovative forms of teaching, as emerges in Chapter 3 on the epistemological nature of geosciences, and in Chapter 4 on the learning process and 'thinking geologically.'

Geosciences, precisely because of their complex nature—from Latin 'cumplexus' (intertwined) and 'cumplicato' (folded together)—offer an ideal ground for promoting basic transversal competencies: systemic thinking, the ability to analyse multidimensional problems, recognising cause-effect relationships in complex systems.

A further aspect concerns the organisation of school curricula themselves. In many national programs worldwide, phenomena like volcanism or earthquakes are presented as separate entities, disconnected from global dynamics and the Theory of Plate Tectonics. This fragmentation, common across different educational systems, limits their attractiveness and interest, preventing students from grasping the complex cause-effect relationships that characterize the Earth System. A central aspect has emerged from research conducted across diverse educational contexts: the need to reorganize methodological approaches, priorities, and even the sequence of the teaching-learning process.

This realization stems from multi-year experience dialoguing with students in their final years of secondary education across various countries. However innovative and engaging, geoscience activities proposed to students who must choose their university path often arrive too late. Their choice is typically made during the penultimate year, when their image of geosciences remains limited to minerals, rocks, volcanoes, and earthquakes—topics that remain a list of names or catastrophic phenomena disconnected from any global context. A fundamental shift in educational approach could change this situation, especially if accompanied by other methodological innovations

1.3 Beyond "stones and disasters": in search of a new approach

To overcome these critical issues and restore to geosciences the role they deserve, a paradigm shift is necessary that highlights their dynamic and technologically advanced nature. Geosciences have never been as relevant as today for perceiving contemporary global challenges: climate change, sustainable resource management, natural hazard prevention and territorial planning are areas where these disciplines play a central role.

The main problem of the traditional approach lies in the fragmented presentation of geological phenomena as separate entities, preventing the grasp of the complex cause-effect relationships that characterize the Earth System. An integrated vision that connects scientific and humanistic disciplines fosters instead to understand how historical events and scientific discoveries have intertwined, making it possible to build unifying models like that of Plate Tectonics.

Educational research has demonstrated how the use of learning objects made with simple materials but designed to stimulate understanding of global dynamics proves effective in promoting interest and comprehension, overcoming the widespread conviction that geosciences are a discipline limited to "rocks and catastrophes." This new approach finds concrete expression in various areas of educational innovation that deserve particular attention.

1.4 Building competencies through innovative approaches

The transformation of geoscience education requires an articulated strategy that integrates various elements: from curricular reorganisation to the use of innovative teaching tools, from territorial development to interdisciplinary integration.

Revolutionizing the curricular structure represents the first central step. As explored in depth in Chapter 5 "Revolutionizing the geoscience curriculum: from Bottom-up to Top-down approach," teachers should move from the traditional bottom-up approach - which starts from minerals and rocks to gradually reach global systems - to a top-down approach that begins with global dynamics and then descends to specific phenomena. By presenting the general framework of Plate Tectonics first, students can better appreciate how individual elements fit into the broader context of Earth systems.

Valorising traditional materials with new perspectives constitutes a second key element. Chapters 6 "Geosciences beyond rocks: models and simulations" and 7 "Minerals and rocks: storytellers" illustrate useful pathways to valorise precisely those "rocks" traditionally considered boring, revealing their fascination and, above all, their relevance to contemporary society. The paleontological record deserves particular attention: Chapter 8 "Fossils as windows on evolution: the paleontological record in education" explores how fossils can become extraordinary educational tools for comprehending not only biological evolution, but also the climatic and environmental changes that have characterized our planet's history.

Reinterpreting catastrophic phenomena in a new interpretative key represents a further innovative aspect. This research proposes a new perspective on Earth Sciences, providing a different interpretation of geological phenomena: natural disasters like volcanoes or earthquakes are not only dramatic phenomena, but causes of events that have profoundly influenced human history. Chapter 9 delves into the theme of natural hazards, proposing a rethinking of the concept of dangerousness beyond NIMBY logic, while Chapter 10 "Catastrophes... and other stories" reveals the surprising connections between geological events and human history.

Integrating territory and technologies offers concrete opportunities for meaningful learning. The study of territory, natural and anthropized environments, offers opportunities for in-depth exploration not only for contact with nature, but also for discovering the geological history and evolutionary dynamics of nearby territories. Chapter 11 "Traditional and digital cartography: developing key competencies through territorial reading" emphasizes the importance of integrating traditional tools and new technologies, while Chapter 12 illustrates how Geosites and Geoparks prove to be extraordinary natural laboratories for geoscience learning, offering the possibility to explore the connection between rocks, minerals, history and society.

Developing specific and transversal competencies through diversified methodologies represents a central objective. Chapter 13 proposes suggestions for building competencies in the geological field, while Chapter 14 "Forensic geology: the detective geologist" illustrates how geological competencies can find application even in investigative contexts, demonstrating the versatility and practical utility of these disciplines in contexts apparently distant from traditional research. Chapter 15 dedicates particular attention to Data Mining Tests as a particularly effective educational tool.

Diversified methodological approaches and the use of different tools - from hands-on artifacts to digital applications - can develop in students not only specific competencies in the geological field, but also transversal competencies of great importance, fundamental in any scientific research and for the integration of soft skills required in the working world. However, as emerges in Chapter 16 "The role of digital technologies in geoscience education," if on one hand digital tools have developed digital competencies and facilitated orientation, on the other hand they have made the use of traditional maps substantially obsolete, causing the loss of observation capabilities and comprehension of territorial complexity.

1.5 Geosciences for sustainable development

Educational innovation in geosciences acquires an even deeper meaning when framed in the context of contemporary global challenges. In an era where environmental issues require urgent and complex responses, geosciences offer not only analytical tools, but a unique perspective that interweaves past, present and future of our planet.

The contribution to global objectives represents an essential dimension of modern geoscientific education. Their importance is recognised by the United Nations 2030 Agenda for Sustainable Development, where numerous Goals require a solid awareness of geological processes and their interactions with human activities. Chapter 17 specifically develops the role of geosciences in achieving the Sustainable Development Goals of the 2030 Agenda, the SDGs.

Interdisciplinary integration emerges as a characterizing element of an innovative educational approach. The interdisciplinary dimension of geosciences manifests with particular clarity when considering their interweaving with history, art and culture. Chapter 18 "Disciplinary interweaving: history, science and art in geosciences" explores how this discipline can become a natural bridge between apparently distant fields, demonstrating how understanding geological phenomena enriches the reading of historical, artistic and cultural processes of humanity.

This integrated perspective not only responds to contemporary educational needs but restores to geosciences that dimension of complexity and fascination that can radically transform students' perception. The hope is that this new approach can ignite that spark of wonder that transforms knowledge into passion and curiosity into commitment: a spark that ignites when students manage to perceive the Earth not as a static ensemble of rocks and distinct phenomena, but as a dynamic and interconnected system of which they are an integral part.

It is the same amazement that has guided the great Earth scientists in their revolutionary discoveries, transforming the observation of a simple rock layer or fossil into an open window on our planet's history and its possible future. Perhaps this can help bridge the gap between the growing demand for geoscientific competencies and the decline of vocations in this fundamental sector, restoring to geosciences the central role they deserve in the education of 21st century citizens.

Namib Desert, dunes in the Sossusvlei area

Further reading

Bruner, J.S. (1960). *The Process of Education.* Harvard University Press. https://www.hup.harvard.edu/catalog.php?isbn=9780674710016

Kastens, K.A., & Manduca, C.A. (2012). *Earth and Mind II: A Synthesis of Research on Thinking and Learning in the Geosciences*. Geological Society of America. https://doi.org/10.1130/SPE486

Strategy for Enhancing Global Geoscience Education – 2019 https://www.iugs.org/_files/ugd/f1fc07_0054e76980a0433897c26aaacddc3742.pdf?index=true

CHAPTER 2.
GEOLOGY, EARTH SCIENCES OR GEOSCIENCES: LET'S TRY TO CLARIFY

In everyday language, in schools, in textbooks, and even in university courses, the terms "geology," "earth sciences," and "geosciences" are often used interchangeably, attributing to them an equivalent meaning: the science that deals with rocks, earthquakes, and natural hazards. However, this simplification hides a complex conceptual evolution that deserves to be explored.

2.1 A question of terminology and substance

The evolution of the terms "geology," "earth sciences," and "geosciences" is not a matter of words but reflects a profound conceptual and methodological transformation in the study of our planet.

Geology, in its classical approach, has focused primarily on the study of rocks, minerals, and the geological processes that have shaped and continue to shape the Earth's crust. This approach, which is still valid, represented the first step toward a systematic perception of our planet.

Over time, this discipline has progressively broadened its horizons towards a more systemic and integrated vision. The definition of "Earth sciences" marked a first significant expansion, including disciplines such as oceanography, meteorology, and climatology, thus recognising the interconnections between the different spheres of the planet.

Modern geosciences are transforming into a high-tech field, far removed from the traditional image - a field geologist with a compass and hammer, engaged in mapping, sample collection, and stratigraphic surveying represents only one aspect of a discipline that today embraces advanced technologies and innovative approaches.

Geophysics uses sophisticated seismic imaging techniques to explore the internal structure of the Earth, while geochemistry and subsoil microbiology are opening new frontiers in the study of interactions between the mineral and biological worlds, with applications ranging from the search for life in extreme environments to the bioremediation of contaminated sites. Geoinformatics and Geographic Information Systems allow us to analyse and visualize complex geospatial data, creating three-dimensional models of the subsurface that are revolutionizing our knowledge of geological structures. At the same time, remote sensing has transformed the way we study the Earth's surface, using drones and LiDAR technologies to accurately map the territory and monitor environmental changes such as deforestation and glacier melt in real time.

The digital revolution has also had a profound Impact on the field of numerical modelling, where supercomputers make it possible to simulate complex geological processes taking place on timescales inaccessible to direct observation, providing increasingly accurate forecasting tools.

Paleoclimatology is particularly important, as it analyses marine sediments and ice cores to reconstruct the climate history of our planet, providing essential data for understanding current climate change. Geoarchaeology, which integrates geophysical and geochemical techniques with traditional archaeological methodologies, is revolutionizing the study of ancient civilizations and their interaction with the environment.

Geoethics, which studies and practices ethical principles in geosciences, addresses the moral and social implications of research and human intervention on the planet. This approach emphasizes the responsibility of geoscientists to promote sustainable practices, ensuring the well-being of future generations through conscious management of natural resources and promoting public awareness of geoscientific issues.

Knowledge of these disciplinary branches, when presented through appropriate teaching processes, can substantially change students' perceptions of geosciences. No longer simply as the study of "rocks and disasters," but as intriguing and vital sciences for awareness and responsibly managing our planet's resources, with a view to sustainability and respect for future generations.

This transformation is reflected in the organisation of universities. While in the Anglo-Saxon world the term "Earth Sciences" still prevails, in Italy we are witnessing a transition from geology departments to broader structures: degree courses in geosciences for the environment, sustainable development, natural resources, and risks are becoming more widespread, while some universities also offer integrated courses in biogeology.

Curricula have evolved from programs focused on the classification of rocks and minerals to structures that reflect an integrated view of Earth processes, including the study of the life cycle of materials, the sustainable management of critical raw materials, and technologies for reducing the carbon footprint.

At the same time, new areas of intervention have emerged, such as the remediation of contaminated sites and the design of environmental restoration measures, which require a multidisciplinary approach and a deep comprehension of ecosystem dynamics.

In schools, however, perhaps due to its similarity to biology, students still think of minerals, rocks, volcanoes, and earthquakes, and when time allows, plate tectonics, as simply "geology."

2.2 Continuity in the education system

This conceptual evolution demands rethinking education systems globally. The adoption of common language between secondary schools and universities is essential to build coherent educational pathways, creating a continuum that accompanies students without conceptual fractures.

Secondary schools worldwide can align their curricula with developments in academia by introducing key concepts such as interdisciplinarity and a systemic approach to Earth phenomena at early stages. Currently, however, many national curricula share similar limitations regardless of country: key topics such as hydrogeological cycles and groundwater dynamics are often reduced to theoretical concepts; local examples of how geological knowledge influences urban planning are rarely integrated into learning processes; connections between geosciences and contemporary challenges such as energy transition, circular economy, and climate change often remain implicit or absent.

While university programs globally are evolving rapidly, integrating advanced topics such as sustainable management of critical raw materials and application of nature-based solutions, these gaps persist at school level across diverse educational systems. This discrepancy creates disconnection in educational pathways that could be bridged by gradually introducing the same principles of sustainability, circular economy, and systemic approach that characterize contemporary university geoscience programs worldwide.

2.3 A look at global geoscience landscapes

Geoscience education varies significantly across countries, reflecting different academic traditions and national priorities. The data paint a worrying picture: according to an IUGS-COGE survey, only 40% of countries have a national curriculum dedicated to Earth Sciences, while 20% explicitly state that they do not. Even more alarming, in about 33% of countries, geosciences are not taught at all in schools, despite being formally included in the curriculum.

In the Anglo-Saxon world, especially in the United States and the United Kingdom, Earth Sciences have traditionally enjoyed greater prestige, with well-funded departments and strong integration between pure and applied research. However, even in these countries, a worrying trend can be observed: data show a significant decline in university enrolments, with the number of geoscience graduates in the United States falling from around 40,000 in 1984 to less than 30,000 in 2020. A similar trend can be seen in the United Kingdom, where first-year enrolments have fallen from over 30,000 in the 1980s to around 10,000 in 2020.

It was recently announced that the Department of Earth Sciences at Vrije Universiteit Amsterdam closes its doors. The decision was announced on April 3, 2025, by the university's management as part of a cost-cutting plan. The measure will result in the loss of over 40 jobs, and the 130 students currently enrolled will be redirected to other programs and universities.

The closure comes at a critical time for geosciences, which are facing a global enrolment crisis. As highlighted in the university's own statement, there is a need to "better communicate the importance of geosciences in order to increase student numbers," at a time when these disciplines are crucial for addressing challenges such as climate change, access to clean water, energy transition, and natural risk mitigation. This case represents the manifestation of a progressive decline, a paradox considering the growing need for geoscience skills to address emerging environmental challenges.

In European countries, the landscape is more varied. The Nordic countries have developed an advanced approach to integration with environmental issues and sustainability. Germany maintains a strong tradition in applied geology and geological engineering, while France has developed a system where the Grandes

Écoles play a key role in training specialists. In Italy, first-year enrolments have fluctuated significantly, but have remained below 2,000 students over the last decade.

The Asian context offers a different perspective. In Japan, geosciences are strongly oriented towards the study of natural hazards and their mitigation, reflecting the country's specific geological challenges. China has invested heavily in recent decades, with a particular focus on mineral resources and remote sensing technologies. However, even in rapidly developing Asian economies, competition from other STEM fields for talented students remains intense, and maintaining robust geoscience programs requires continuous effort.

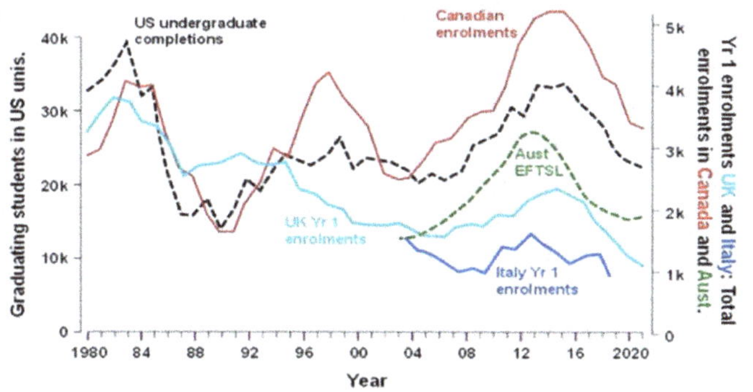

Comparisons of earth science enrolment trends in the US, UK, Canada, Italy and Australia (Cohen, 2022; Bonaccorsi et al., 2020; CCCESD, 2020)

Latin American and African contexts face additional challenges beyond enrolment declines. Institutions in these regions often struggle with limited research funding, outdated equipment, and difficulty retaining trained faculty who frequently migrate to better-resourced institutions elsewhere. This creates a troubling paradox: regions extraordinarily rich in geological resources and facing significant natural hazards often possess the weakest geoscience education infrastructure.

A striking finding from the surveys is that only 33.3% of countries offer specialized courses on 'how to teach' Earth Sciences, highlighting a significant gap in teacher training in this discipline.

Data on the distribution of graduates in different geoscience specializations reveal further significant trends: traditional Earth Sciences account for only 7.5% of graduates, followed by Environmental Sciences (4.8%) and Environmental Geosciences (3.4%). Specializations such as Hydrology and Water Resources

(0.8%) or Petroleum Geology (0.3%), despite their relevance to global challenges, attract a minimal percentage of students.

In this global landscape, some common trends are emerging increased interdisciplinarity, a growing emphasis on digital technologies and big data, and a focus on global challenges. However, significant differences persist in the organisation of educational programs and teaching methodologies.

2.4 Towards a new disciplinary identity

The terminological and conceptual transformation of geosciences is leading to a profound redefinition of their identity in the 21st century. Scientific institutions and research bodies are renewing not only their names but also their internal structures, reflecting the transition from a predominantly descriptive science to an interpretative and predictive discipline.

The interdisciplinary nature of geosciences goes far beyond its relationships with other natural sciences. Historically, geological phenomena have influenced entire civilizations: ancient societies prospered or declined in conjunction with climate changes caused by geological processes; economic development has always been linked to mineral resources; population movements have often followed the geological conformation of the territory.

The new identity of geosciences is based on three main pillars:

1. **The centrality of systemic thinking**: modern geologists do not simply catalogue rocks and minerals, but interpret complex systems and their interactions, reading environment as a set of interconnected processes that require a comprehensive view
2. **A focus on solving concrete problems**: from theoretical research, geosciences have become a strategic discipline for addressing environmental, energy, and territorial issues.
3. **Social responsibility**: the scientific community recognises

This new disciplinary identity requires a rethinking of the educational approach, with the objective of developing in future geoscientists not only technical knowledge, but also critical thinking abilities, communication skills and a solid awareness of the social role of their discipline. In the following chapters we will explore how this transformation implies new educational methodologies to build the foundations of scientific culture for 21st century citizens.

Grand Prismatic Spring –Yellowstone Park

Further reading

Bonaccorsi, E., Occhipinti, S., Borghini, A., & Greco, R. (2020). "Student Enrolment in Geology from a Systemic Earth Science Education Perspective: An Italian Case Study." *European Geologist*, 50. https://eurogeologists.eu/journal/student-enrolment-in-geology-from-a-systemic-earth-science-education-perspective/

Francek, M. (2013). "A Compilation and Review of over 500 Geoscience Misconceptions." *International Journal of Science Education*, 35(1), 31-64. https://doi.org/10.1080/09500693.2012.736644

Van Loon, A.J. (2008). "Geological education of the future." *Earth-Science Reviews*, 86(1-4), 247-254. https://doi.org/10.1016/j.earscirev.2007.08.005

CHAPTER 3.
THE NATURE OF GEOSCIENCES: COMPLEXITY AND SYSTEMS, FLOWS AND RELATIONSHIPS

Analysing the characteristics of geosciences, particularly the nature of knowledge and the processes through which it is acquired and understood, reveals the need to develop a more solid epistemological framework for this discipline. This need becomes even more evident when examining the historical evolution and fragmentation that has characterized geosciences.

3.1 The importance of epistemology for geosciences

Epistemology, i.e., the study of the nature, origin, and limits of knowledge, is essential to interpret how knowledge is constructed in any scientific field. All scientific disciplines are based on an epistemology that allows us to examine the foundations of the discipline itself, the conditions that allow scientific knowledge to be constructed, and the methods for achieving it.

In the case of geosciences, this reflection is especially relevant due to the specificity of the phenomena studied: processes developing on temporal and spatial scales that transcend not only direct observation but often also human intuitive awareness. The geological "deep time," measured in millions or billions of years, and the dynamics that have shaped our planet require a different epistemological approach than disciplines where phenomena are more easily observable and reproducible.

It is no coincidence that epistemological reflections have mainly involved biologists, physicists, mathematicians, and astronomers, and rarely geologists. This pattern holds true across different scientific traditions worldwide, though the specific historical development of epistemological thought about Earth sciences has varied by region. European traditions emphasized natural philosophy, while North American approaches integrated more pragmatic, resource-oriented perspectives, and Asian traditions often maintained stronger connections between Earth studies and holistic worldviews.

The latest generations of geoscientists have been engaged in other directions: raw materials, geological hazards, ocean exploration, geophysical investigations, and, of course, investigations into the history of the Earth in all its aspects. It is natural to think that the lack of unity in the discipline is one of the causes of this lack of interest in theoretical aspects.

3.2 Fragmentation of the discipline and historical evolution

Although geosciences are relatively young as a formal discipline, their roots can be traced back to the pioneering work of scientists such as Steno, Hutton, Lyell, and Argand in the 18th and 19th centuries. These scholars laid the foundations for modern geology, developing central principles and reconstructing the geological history of the Earth. However, unlike other sciences that early on developed a coherent model of knowledge, the geosciences followed a more fragmented path.

At the beginning of the 20th century, the concept of the Earth as a system, resulting from the interactions of its various components, began to take shape, but a unifying theory was slow to emerge. Throughout the century, numerous scientists focused on their individual fields, working with a reductionist approach, rather than building the interconnected knowledge necessary to develop a comprehensive theory of the discipline.

A characteristic phenomenon of this period was the progressive separation between different specializations: structural geologists ceased to communicate with volcanologists and geophysicists, geographers with geomorphologists, sedimentologists diverged from stratigraphers, micropalaeontologists from palaeontologists, mineralogists from petrographers, while applied geologists distanced themselves from naturalist researchers.

It took almost 70 years for the theory of plate tectonics to emerge, which, although it provided a unified explanation for many of the major phenomena on Earth, failed to completely unify the fragmented field of Earth sciences. The discipline remains divided into various branches, each dealing with different environments—land, air, water, space, soil, subsoil—but all closely interconnected. This historical fragmentation has hindered the development of a unified epistemology.

Several scholars have reflected on these issues, such as Hubral, Tarantola, Baker, and Oldroyd, but relatively few philosophers of science have addressed this discipline, partly because of its atypical characteristics, which make it "outside the box" of traditional philosophical paradigms of science.

Among the most important contributions is that of Victor Baker, who proposed the concept of "geosemiotics": a way of doing geology that combines the interpretation of "signs"—rocks, structures, and landscapes—with the application of physical and chemical laws to terrestrial processes. This approach remains at the heart of geology as a science and profession, even if modern technologies seem to overshadow direct observation.

Other important scholars include David Oldroyd, who in "Thinking About the Earth" (1996) studied the evolution of geological thought; Robert Frodeman, who in "Geo-Logic" (2003) showed how geosciences have their own specific way of reasoning, combining deduction and interpretation; Henry Frankel, with "The Continental Drift Controversy" (2012); Martin Rudwick, with Earth's Deep History (2014); and Naomi Oreskes, with The Rejection of Continental Drift (1999). According to Iannace (2023), the birth of geology was the result of the union of two approaches: the historical-Neptunist and the causal-plutonist. These approaches merged thanks to the collaboration of many European scientists when the Geological Society of London was founded in 1807, an event that can be considered the official birth of this science.

Despite these unifying efforts, contemporary studies still highlight the need for a more unified and coherent epistemological approach that can account for the specificity of geosciences and their peculiar way of constructing knowledge.

3.3 Geoscience as a 'non-Galilean' discipline

Geosciences are characterized by an exceptional scope in both space and time, embracing the perception of deep space, from the Earth's core to the boundaries of the Solar System, and the grasping of deep time, from the formation of the Solar System to the present day. These distinctive features require specific methodological approaches to address the unique challenges posed by the study of the Earth.

One of the most significant aspects of the geoscience methodology is the impossibility of fully applying the Galilean experimental method. Geological processes, which develop over timescales of millions of years and on immense spatial scales, make direct observation and experimental replication impossible.

Experimentally replicating an orogeny that develops over tens of millions of years or reproducing the subduction of an oceanic plate in the laboratory are phenomena that transcend not only the technical possibilities of experimentation, but also the timescales of human observation.

It is universally known that the typical way in which science proceeds to achieve objective, reliable, verifiable, and shareable knowledge of reality is the scientific method, or experimental method, introduced by Galileo in the 17th century. It consists of collecting empirical data based on the formulation of hypotheses and the rigorous, rational, and, when possible, mathematical analysis of these data. Every phenomenon must be verifiable, observable, and repeatable; every variable introduces variations, which in turn can be formulated through mathematical functions.

This paradigm, which has enabled the development of disciplines such as physics and chemistry, has major limitations in the geosciences. The difficulty in applying the Galilean experimental method has led some to question the scientific status of geosciences, arguing that if the traditional scientific method cannot be applied, then “it is not science”. But this is a mistake in perspective: it is not the discipline that lacks scientific rigor, but the traditional method that is inadequate for appreciation the complexity of geological phenomena.

Faced with these difficulties, geosciences have developed an alternative approach based on the construction of interpretative models that integrate partial observations, indirect data, and theoretical inferences. Far from being less rigorous, this approach represents an epistemological richness that characterizes the discipline and contributes to its scientific identity.

3.4. The role of interpretation in geological knowledge

A key aspect of the epistemology of geosciences is the central role of interpretation. As highlighted by philosopher Thomas Kuhn in his work "The Structure of Scientific Revolutions" (1962), plate tectonics is "a profoundly revolutionary theory - a new paradigm because it has forced us to rethink almost all the major themes of geology, from tectonics to orogenesis to biogeography". In his view, geology requires a strong interpretative component in reconstructing Earth's history from fragmentary evidence.

This approach emerges clearly in stratigraphy and Palaeontology, where the reconstruction of the past requires not only objective data, but also interpretative capabilities based on theoretical frameworks and analogies with current processes. Geosciences are therefore characterized by a constant balance between reductionism and holism, between detailed analysis of individual components and the need to consider the Earth system as a whole. This methodological polarity reflects the specific complexity of geological systems and the necessity for multi-scale approaches.

The interpretative approach of geosciences recognises that terrestrial phenomena are characterized by a network of interactions that transcends simple linear causality, requiring a rethinking of traditional epistemological paradigms. The circular and non-linear connections between components are so interconnected that they cannot be separated, physically and conceptually, without compromising the entire system. Consequently, the analytical method typical of classical science, based on decomposition into independent modules, may prove inadequate for mastery a complex system. The principle of uniformitarianism - or actualism -, fundamental in geology, clashes with the incompleteness of stratigraphic and fossil records in deep time, making direct observation impossible and creating theoretical fragilities.

This interpretative approach extends also to the interdisciplinarity that characterizes modern geosciences. On one hand, there persists a strong value of observation and description, inherited from the naturalistic tradition; on the other, a quantitative and experimental approach has developed, influenced by the physical sciences. The integration of these approaches is not simple but represents a unique epistemological richness. The study of volcanic eruptions perfectly exemplifies this integration: direct observation of eruptive phenomena combines with precision geochemical analyses and complex numerical models, creating an interdisciplinary framework that facilitates comprehension and, in part, predicting the behaviours of volcanic systems.

3.5 Complexity as a contribution to a geoscience epistemology

In formulating an epistemology appropriate to geosciences, three central concepts must be placed at the centre: complexity, the system, and the interconnected relationships that characterize Earth processes.

Complexity in geosciences refers to a set of intertwined elements, where significant properties arise from interactions between components and cannot be predicted by isolated analysis of individual parts. The is a perfect example of this characteristic: biogeochemical cycles, lithosphere-hydrosphere-atmosphere-biosphere interactions, and global tectonic dynamics give rise to phenomena that exceed the simple sum of their individual components.

Each scientific discipline has its own dimension to study, which differs in terms of the field of study and the variety of techniques used, but traditional sciences tend to be reductionist and are therefore inadequate for studying complexity. The study of geosciences, on the other hand, offers a unique opportunity to conceptualize phenomena across time and space, in a complex and interactive historical approach that goes beyond the purely analytical and reductionist approach typical of other scientific disciplines. An epistemology of geosciences must recognise that to interpret geological phenomena requires both analysis and synthesis, the ability to break things down and put them back together, to see the particular in the general and the general in the particular. This means learning to use different scales of observation simultaneously, from the microscopic (crystals, microfossils) to the macroscopic (landscapes, mountain ranges), up to the planetary scale of global processes.

The concept of the system is equally essential and represents perhaps the most significant contribution that geosciences can offer to science education. The Earth System consists of interconnected subsystems that interact continuously through exchanges of matter and energy. These exchanges are not linear but cyclical, characterized by positive and negative feedback loops that regulate the dynamic equilibrium of the planet. Grasping the concept of systems means developing the ability to identify connections, often not immediately apparent, between phenomena that are apparently distant in space and time.

Relationships and flows constitute the third pillar of this epistemology. In geosciences, nothing exists in isolation: every rock tells the story of the conditions in which it was formed, every landscape is the result of the interaction between endogenous and exogenous processes, every geological event has repercussions on multiple components of the system. The flows of matter and energy that traverse the Earth System—from the rock cycle to biogeochemical cycles, from ocean currents to convective currents in the mantle—represent the "connectors" that bind the different parts of the system together.

This epistemology of geosciences centred on complexity, systems, and relationships is not just a theoretical conceptual model, but has important consequences.

It allows us to overcome the fragmentation of geological knowledge by linking seemingly distant phenomena into a coherent framework. It makes it possible to understand processes that take place on spatial and temporal scales that exceed direct human experience. It provides not only disciplinary knowledge, but also real tools for interpreting the world, recognising the interconnections between natural phenomena and human activities, and recognise our role within this System.

Above all, it fosters the development of a form of thinking—systemic thinking—that is essential not only for understanding geosciences, but also for addressing the environmental and social challenges of the 21st century. According to complexity theory, scientific representations of reality are due to phenomena and problems immersed in a system that encompasses them, which we call the environment. This view, which may seem abstract, has important consequences for how we study and teach geosciences, as we will see in the following chapters on the learning process and teaching.

Columnar basalt; Iceland Fljotsdalsherao

Further reading

Alvarez, W. (2017). *A Most Improbable Journey: A Big History of Our Planet and Ourselves*. W.W. Norton & Company. https://wwnorton.com/books/9780393355192

Frodeman, R. (2014). *Geo-Logic: Breaking Ground Between Philosophy and the Earth Sciences*. SUNY Press. https://www.sunypress.edu/p-5846-geo-logic.aspx

Kuhn, T. (1962). The Structure of Scientific Revolutions. University of Chicago Press. https://www.einaudi.it/catalogo-libri/filosofia/filosofia-contemporanea/la-struttura-delle-rivoluzioni-scientifiche-thomas-s-kuhn-9788806229757/

Stewart, I.S., & Lewis, D. (2017). "Communicating Contested Geoscience to the Public: Moving from 'Matters of Fact' to 'Matters of Concern'." *Earth-Science Reviews*, 174, 122-133. https://doi.org/10.1016/j.earscirev.2017.09.003

CHAPTER 4.
THE LEARNING PROCESS IN GEOSCIENCES: CAN WE "THINK GEOLOGICALLY"?

The learning process in geoscience represents a unique cognitive journey that differs significantly from other scientific disciplines. Recognise how students acquire and process knowledge in this field requires careful consideration of fundamental cognitive mechanisms, motivational factors, and specific learning challenges.

4.1 The transition from spontaneous curiosity to structured learning

Learning is an innate tendency in developing organisms, particularly evident in the way younger students, such as primary school pupils, approach the natural world. Their endless questions about natural phenomena and adventurous exploration of their environment demonstrate a desire to know and understand. However, this spontaneous curiosity often diminishes as they grow older and are confronted with formal, structured curriculum content.

While younger children are fascinated by spectacular phenomena such as volcanic eruptions or iconic elements such as dinosaur fossils, which are capable of arousing wonder regardless of the geological or evolutionary context in which they are presented, their engagement inevitably tends to diminish when they are confronted with more complex concepts, such as the internal structure of the Earth or the genesis of rocks.

The cognitive processes involved in learning geosciences are complex and complicated: they are therefore intertwined and interconnected. Students need to develop the ability to integrate new information with existing knowledge, construct mental schemas for abstract concepts, and recognise patterns and relationships in the functioning of Earth systems. This task requires both understanding individual concepts and developing strategies to monitor one's comprehension—knowing that one understands—by adapting study methods to the complexity of the topic. As natural curiosity transforms into formal learning, motivation becomes increasingly important. Students need to have control over their learning process, be aware of their progress, and experience the growth of their abilities through progressive challenges. It is essential that they connect their learning to personal interests and real-world applications, thus completing a virtuous cycle where metacognition facilitates them not only to acquire knowledge but also to assimilate it deeply, recognising its importance and practical value in their daily lives.

4.2 Cognitive complexity and multidimensional thinking

Learning geosciences involves cognitive processes that require highly developed abstraction and visualization skills. Unlike other experimental sciences, geology requires an awareness of time scales that transcend human experience, from the instant of an earthquake to the millions of years required for the formation of mountain ranges.

'Deep time' represents one of the most significant cognitive challenges. The human brain, evolutionarily programmed to understand time scales related to direct experience, struggles to mentally represent processes that take place over millions or billions of years. Students tend to compress geologically distant events or misperceive phenomena separated by enormous time intervals as contemporary. Equally challenging is the spatial visualization required by geosciences, where students must be able to imagine processes occurring on scales ranging from microscopic crystal lattices to global movements of tectonic plates. This ability to mentally "zoom" from the microscopic to the planetary, while maintaining awareness of the causal relationships between phenomena operating on completely different scales, represents a considerable cognitive challenge.

The ability to visualize three-dimensional structures from two-dimensional representations is also significantly complex. Interpreting geological maps, cross-sections, and underground structures requires advanced spatial visualization skills, using specific areas of the brain. Students must learn to mentally reconstruct shapes and volumes from partial information, a skill that goes far beyond the simple memorization of facts and definitions.

When students acquire these complex cognitive skills, the areas of the brain associated with motivation and reward are activated, creating a virtuous circle of learning that can fuel a spontaneous interest in the discipline.

4.3 Interdisciplinary integration and systemic vision

Understanding geological phenomena requires analysis of the relationships between the various components of the Earth system, which requires integration with other disciplines. The reductionist approach, which has characterized much of the history of Western science, has clear limitations when applied to geosciences. A sedimentary basin or a volcanic system cannot be understood by studying their components separately but require analysis of the complex interactions between these elements in time and space.

Integration with chemistry is necessary to recognise the composition and transformations of minerals; links with physics are essential for the study of seismic wave propagation and fluid mechanics; the connection with biology is crucial for understanding fossilization processes and biogeochemical cycles; and the contribution of mathematics and computer science is indispensable for modelling complex systems.

The most effective teaching approaches clearly highlight these interdisciplinary connections, helping students to overcome the compartmentalization of knowledge. An example is the integrated study of a river basin, which requires simultaneous knowledge of geomorphology, fluid physics, water chemistry, ecology, and mathematical modelling.

The systemic approach requires overcoming linear cause-and-effect thinking in favor of interpret nonlinear dynamics. In geosciences, small causes can produce large effects, or vice versa. We must learn to understand complex systems, where small causes can have large effects, where effects arrive late, and where reactions are often unexpected.

It is also important to identify the positive and negative feedback loops that characterize Earth systems, from tectonics to ecosystems. Understanding these loops requires the ability to think in terms of cyclical rather than linear processes, visualizing connections that span different spheres of the Earth system.

4.4 Rethinking the structure of geological knowledge acquisition

In geosciences, the traditional structure of knowledge acquisition often follows a predetermined path influenced by curricula, textbooks, and established teaching approaches. The conventional path typically proceeds from simple to complex, in a sequence that begins with mineralogical and petrographic fundamentals, focusing on the identification of minerals and rocks as the basic units of geology.

Students learn to recognise and classify these key elements before moving on to more complex levels. Subsequently, the focus shifts to geological and geomorphological processes, such as volcanoes, earthquakes, and river systems, which are initially analysed as individual elements. Only in the most advanced stages are regional systems and global dynamics, such as plate tectonics, introduced.

Although seemingly logical, this structure has significant drawbacks: it fragments knowledge into separate compartments, delays the perception of systemic connections, and reduces motivation by focusing initially on technical aspects before students can understand the big picture. An alternative approach that reverses this traditional sequence, starting with global dynamics and gradually moving down to the details, could provide a comprehensive interpretative framework from the outset.

This traditional sequence, common in curricula worldwide from North America to Europe to Asia, has been questioned by educators across diverse educational contexts who observe similar patterns of student disengagement and fragmented knowledge regardless of cultural or linguistic differences.

However, these theoretical principles must be translated into concrete teaching approaches through a reorganisation of teaching practices that responds to the cognitive and epistemological peculiarities of geosciences, suggesting methodologies that facilitate the transition from the mere acquisition of knowledge to the development of authentic 'geological thinking'.

4.5 Teaching strategies for the development of geological thinking

The teaching of geosciences requires a diversification of educational methodologies to respond to students' different learning modalities and the specific cognitive challenges of the discipline.

Students who learn visually benefit most from diagrams, maps and 3D models. Technologies like augmented reality allow them to 'see' otherwise inaccessible phenomena, transforming the learning experience.

Many students better assimilate information through stories and verbal explanations that connect geological events in narrative sequences, building stories about the planet's history. These students benefit from discussions, podcasts, audio guides and dialogic lessons where concepts are elaborated orally. For them, transforming complex geological processes into coherent narratives greatly facilitates comprehension and memorization. Others, instead, learn better through practice and manipulative activity, benefiting from direct experience: fieldwork, manipulation of rock samples and creation of three-dimensional specimens. The "hands-on" approach supports them to develop a concrete appreciation of geological processes, connecting physical sensations to abstract concepts.

Effective teaching must integrate methodologies that satisfy all these styles, offering a diversified repertoire of learning experiences and recognising that deep understanding of geosciences requires the integration of different cognitive modalities. Modern educational technologies offer unprecedented opportunities for this personalization, allowing students to interact with geological content according to their own preferences.

Inquiry-Based and problem-based learning

Inquiry and investigation-based learning is notably suited to geosciences, as it reflects the real process through which geological knowledge is constructed. In this method, students start from authentic questions about geological phenomena, formulate hypotheses based on prior knowledge, design and conduct investigations to test their hypotheses, analyse collected data and draw conclusions which they then communicate and discuss with peers.

This method develops investigative skills and critical thinking, placing the student at the center of the learning process. Exceptionally effective is the use of real case studies, where students must apply the scientific method to solve concrete geological problems.

Problem-Based Learning or Problem-Solving presents students with real geological challenges such as predicting landslide risk in a specific area, identifying the optimal site for a water collection facility, or reconstructing the geological history of a territory starting from limited outcrops.

Educational experience demonstrates that when students tackle open problems like "Determining the hydrogeological risk of their own town," they begin to think like geologists: they integrate diverse knowledge, propose hypotheses and defend them with concrete evidence. Gradually they move from a fragmented vision to a systemic understanding, learning to connect apparently disconnected elements.

This methodology is particularly effective because it forces students to face open problems, for which there is no single correct solution, but different possible interpretations that must be evaluated based on available evidence - exactly as occurs in real geological practice.

Experiential learning and digital storytelling

Field experiences are an irreplaceable element of geological education. Direct observation of rock outcrops, geological structures, and active processes in the field provides an awareness that no textbook or lecture can offer.

Field activities include excursions to geologically significant sites, mapping local geological structures, sampling and analysing rocks and minerals, and monitoring active geological phenomena such as erosion or sedimentation. These direct experiences develop practical skills, observational abilities, and adaptability in variable natural contexts.

Fieldwork also allows students to develop what we might call "landscape literacy": the ability to read the land like an open book, recognising in the forms of the landscape the traces of the geological processes that have shaped it over time.

Digital storytelling can be used effectively to communicate "geological stories," making complex concepts accessible through engaging narratives. This approach can include reconstructing the evolution of a landscape over geological time, narrating the "biography" of a mineral from its formation to its use by humans, or presenting a geological investigation as a "mystery" with clues and solutions.

Digital storytelling makes possible the integration of text, images, video, animation, and sound to create immersive, multisensory learning experiences. This approach is notably effective for communicating complex concepts such as geological time or tectonic processes, which can be difficult to visualize through traditional methods.

The detective geologist: an integrated educational model

The "detective geologist" approach is a powerful educational model that transforms the teaching of geosciences from the transmission of knowledge to the development of cross-cutting skills. This enables students to become active participants in the learning process, formulating hypotheses and seeking evidence to confirm or refute them, just like a professional geologist in the field.

Geological phenomena are thus contextualized within a network of relationships with other disciplines, from physics to biology, history to economics, allowing students to understand how geosciences are not an isolated field, but a discipline deeply interconnected with all aspects of human knowledge.

The skills developed through this method transcend the specific context of geosciences, preparing students to tackle complex issues in many fields. The ability to analyse complex systems, integrate data from different sources, and think on multiple spatial and temporal scales are valuable skills in many professional and scientific fields.

The fascination of investigation and discovery also stimulates intellectual curiosity and motivates further study, transforming the study of geosciences into a fascinating intellectual adventure. Just as Sherlock Holmes used his deductive method to solve criminal cases from seemingly insignificant clues, geologists read the clues hidden in rocks, minerals, and fossils to unravel the mysteries of the Earth and understand the complex interconnections that bind our planet, its history, and our future.

Metacognition and contextualization in geological learning

Metacognitive skills—the awareness and management of one's own learning processes—are essential for dealing with the complexity of geological phenomena. The proposed teaching models promote the development of these skills by encouraging students to plan their investigative approach, monitor their understanding, and critically evaluate their interpretations. The 'conceptual obstacles' typical of geosciences, such as comprehending geological time or visualizing invisible structures in three dimensions, become opportunities to develop targeted metacognitive strategies.

The contextualization of geological knowledge is a central element of these teaching approaches. The study of geological phenomena in the local area creates immediate connections with everyday experience, transforming abstract concepts into observable realities. Analysing the influence of geological events on human history connects the natural sciences with the humanities, while examining the geological implications for contemporary challenges such as climate change or resource management highlights the social relevance of the discipline. Exploring the connections between geology and everyday life—from minerals in smartphones to the characteristics of drinking water—transforms geology from an abstract discipline into a key to interpreting everyday reality, thus completing the journey toward "thinking geologically."

4.6 'Thinking geologically' as a key to interpreting the world

'Thinking geologically' means overcoming a fragmented view and adopting a systemic approach. Students must learn to recognise the connections between different phenomena, understand complex dynamics, and think on different scales of space and time. Teaching must evolve from the transmission of knowledge to the development of cognitive skills, from the memorization of facts to the mastery of processes, from a static view to a dynamic perception of Earth phenomena.

Thinking geologically is not only a disciplinary goal, but a way of interpreting reality that shapes informed citizens.

Geyser in Yellowstone National Park- USA

Further reading

Kastens, K.A., & Manduca, C.A. (2012). *Earth and Mind II: A Synthesis of Research on Thinking and Learning in the Geosciences*. Geological Society of America. https://doi.org/10.1130/SPE486

Occhipinti, S. (2020). "Motivation in Learning Geosciences. The Teaching-Learning Process: Analysis, Evaluations, and Proposals." DOI: 10.5772/INTECHOPEN.1007983. https://www.intechopen.com/chapters/70778

McLaughlin, J. A., & Bailey, J. M. (2022). Students need more practice with spatial thinking in geoscience education: a systematic review of the literature. Studies in Science Education, 59(2), 147–204. https://doi.org/10.1080/03057267.2022.2029305https://link.springer.com/chapter/10.1007/1-4020-3613-2_13

FROM THEORY TO PRACTICE

In the previous chapters, the theoretical foundations of educational innovation and the evolution of learning processes have been examined; educational models have been proposed that can respond to contemporary challenges and the needs of 21st century students. These theoretical reflections must now translate into concrete educational practices.

The second part of this book aims to bridge the gap between theory and practice, offering concrete examples, tools and methodologies, with the objective of transforming education considering new knowledge about learning and adopting more effective teaching approaches without losing scientific rigor.

The transition from theory to practice is never simple nor straightforward. It requires creativity, experimentation and a deep understanding of both content and learning processes. The following chapters will illustrate this journey through concrete examples of curricular transformation, taking geosciences as a case of a discipline that needs to be rethought in its educational approaches.

CHAPTER 5.
REVOLUTIONIZING THE GEOSCIENCE CURRICULUM: FROM A BOTTOM-UP TO A TOP-DOWN APPROACH

Geoscience education is at a critical juncture: while its importance for addressing sustainability challenges is widely recognised, the discipline is still viewed as fragmented and marginal.

The journey from theory to practice begins by exploring how to revolutionize geoscience education through a radical rethinking of the teaching approach, moving from a fragmented, bottom-up vision to an integrated, top-down vision that reflects the systemic and interdisciplinary nature of this fascinating discipline, a revolution in the curriculum representing a model that could potentially be applied to other fields of knowledge.

5.1 The limits of the traditional bottom-up approach

In geosciences, the traditional structure of knowledge acquisition typically follows a 'bottom-up' path, proceeding from the simple to the complex. This progression begins with the fundamentals of mineralogy and petrography, focusing on the identification of minerals and rocks as basic units. Only later are local geomorphological processes, biogeochemical cycles and the climate system, regional systems and finally global dynamics represented by plate tectonics introduced, divided by school years and not always in continuity.

Although seemingly logical from the point of view of the progressive construction of knowledge, this structure presents important educational problems. The bottom-up approach tends to fragment knowledge into separate compartments, creating the impression that the various aspects of geology are disconnected from each other rather than parts of an integrated system. Furthermore, it significantly delays the knowledge of the systemic interconnections between different phenomena.

Perhaps the most problematic aspect concerns motivation: by focusing initially on technical aspects such as mineral classification, this approach risks significantly reducing the initial interest of students confronted with specialized details before understanding the big picture and the relevance of what they are studying.

The traditional sequence also presents a major problem for orientation: university choices are made towards the end of high school, when the image of geosciences is still limited to minerals, rocks, volcanoes, and earthquakes, elements often presented as isolated phenomena, disconnected from a systemic context.

5.2 The top-down approach: a conceptual revolution

An effective transformation of geoscience teaching could be represented by a shift from the traditional 'bottom-up' approach to a 'top-down' approach, which starts with global dynamics and then progressively moves down to specific phenomena.

This methodology has numerous advantages. Starting from plate tectonics as an essential interpretative framework, students immediately gain an overview of the processes that shape our planet. This global perception provides a meaningful context for the subsequent study of more specific phenomena.

Phenomena such as volcanism, seismicity, or mountain formation are thus perceived not as isolated events but as manifestations of interconnected planetary dynamics. Student motivation remains high because they understand from the outset the relevance and interconnections of what they are studying.

When the focus later shifts to minerals and rocks, these no longer appear as simple objects to be classified, but as tangible evidence of fundamental geological processes, significantly increasing interest and deep awareness. In this way, it is possible to avoid the fragmentation typical of traditional teaching, creating from the outset an interconnected conceptual network within which to place specific knowledge.

The top-down approach is not simply a teaching innovation but represents an approach more in line with the nature of modern geosciences. Geosciences are characterized by their systemic nature, the interconnection of phenomena, and the predominance of relationships over isolated entities: the traditional bottom-up approach, initially focused on discrete entities such as minerals and rocks, reflects a reductionist view that does not correspond to the epistemological evolution of the discipline.

The top-down model, on the other hand, recognises that geological phenomena can only be understood if their interconnections are recognised. When a geoscientist examines a rock outcrop or a mineral, they never do so as an isolated object, but always as part of a larger system and as evidence of processes operating at different scales. In this sense, the top-down approach is not only more effective from an educational point of view, but also more authentic in representing how contemporary geoscientific thinking works and how knowledge is constructed in this field.

5.3 Practical implementation of the top-down approach

A top-down geoscience curriculum unfolds through successive phases.

The process begins by presenting plate tectonics as a unifying interpretative key for understanding geological phenomena. This allows students to familiarize themselves from the outset with the theoretical framework that explains and interconnects different terrestrial processes.

Volcanoes, earthquakes, mountain formation, oceans and sedimentary basins are then introduced as direct manifestations of plate movements, highlighting the causal relationships between planetary dynamics and local phenomena.

The approach is enriched by exploring the interactions between Earth systems: atmosphere, hydrosphere, biosphere and geosphere interact in response to tectonic processes, creating feedback cycles that regulate the Earth system as a whole.

Only after having built this solid conceptual framework does attention shift to details: the study of rocks and minerals now assumes a completely different meaning, since these elements are contextualized as products and witnesses of the great planetary processes previously illustrated.

Finally, local geological phenomena become privileged windows for become aware of global dynamics. When presented as specific manifestations of planetary processes, rather than as isolated cases, they acquire a deeper and more stimulating meaning for students.

This approach has been successfully applied in various educational contexts, showing that students who begin with a global perception of the Earth system maintain a higher level of motivation during subsequent study of details, perceiving them as meaningful parts of a larger system rather than as isolated notions.

5.4 Integration with the spiral learning model

The top-down approach can be effectively combined with the "spiral" curriculum model, in which fundamental concepts are cyclically reintroduced at increasing levels of complexity. This pedagogical approach, theorized by Jerome Bruner in the context of American educational reform in the 1960s and subsequently adopted and adapted in educational systems worldwide, proves particularly suited to geosciences, offering a framework that respects the interconnected and multiscale nature of geological phenomena regardless of specific national curriculum structures.

In the spiral model, key concepts are not abandoned once introduced but are continuously revisited in progressively more complex and sophisticated forms. This approach recognises that deep perception of geological concepts requires multiple exposures and reworkings, each of which adds new levels of complexity and meaning.

The implementation of the spiral approach in geosciences follows a cyclical and progressive path. The first exposure introduces basic concepts in simplified form, building the necessary cognitive foundations. Each new cycle deepens the same themes with increasing complexity, revealing nuances and variations not previously considered. The process is enriched by progressively connecting geological concepts to other disciplines, ultimately transforming theoretical knowledge into practical competencies applicable in real contexts.

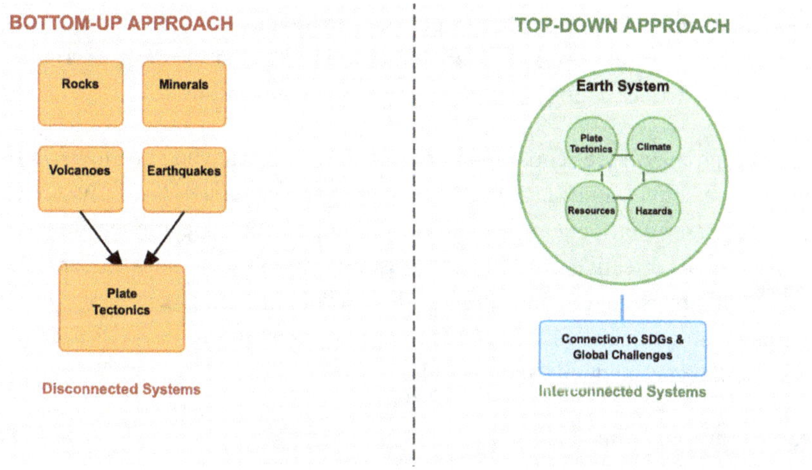

For example, the concept of Plate Tectonics can be initially introduced as a simple model of rigid plates moving on the Earth's surface. In subsequent revisitations, one can explore the physics of plate movement, the complex interactions at margins, the implications for climate and biodiversity, up to applying the knowledge to understand geological hazards or the distribution of mineral resources.

5.5 Addressing implementation challenges

The adoption of a top-down approach to geoscience education, while pedagogically sound, may encounter practical obstacles that deserve careful consideration. The transition from an established teaching model to an innovative one requires not only theoretical conviction, but also concrete strategies to overcome inevitable resistance and operational difficulties.

Formal curriculum constraints

Educational authorities and curriculum frameworks in most countries still follow the traditional bottom-up sequence. This presents a complex challenge, as teachers worldwide often find themselves navigating between official guidelines—which in many educational systems serve as frameworks rather than rigid prescriptions—and approaches they believe to be most effective pedagogically. The degree of curricular flexibility varies significantly by country: some systems allow considerable teacher autonomy in sequencing topics, while others maintain more prescriptive structures. However, across diverse contexts, it remains possible to develop adaptation strategies that respect institutional requirements while modifying their practical implementation.

The solution may lie in reorganizing the timing of topics while keeping the final learning objectives unchanged. A gradual approach could prove particularly effective: start by introducing the top-down model within specific teaching units, demonstrate its effectiveness through measurable results, and only then propose a more comprehensive reorganisation of the curriculum. This incremental process facilitates the gradual building of consensus and trust in the new approach, both among colleagues and with school authorities.

Teacher preparation

Most teachers have received training based on the traditional approach and may not feel adequately prepared to implement a different paradigm. This insecurity is understandable and deserves to be addressed seriously. It is not simply a matter of explaining a new approach in theory, but of accompanying teachers through a real process of professional transformation.

Specific training courses that go beyond simply illustrating the top-down method and instead offer direct experience of curriculum design according to this paradigm are essential. Practical workshops, in which teachers can personally experience the construction of top-down teaching units, are really effective. Equally important is the creation of communities of practice where teachers can share experiences, exchange materials, and discuss solutions adopted to overcome specific difficulties.

Teaching resources

A significant practical obstacle is that most textbooks and teaching materials available still follow the traditional approach. Teachers who wish to adopt the top-down approach often find themselves having to create or adapt materials themselves, which requires a considerable investment of time and energy.

To address this problem, teaching materials designed for the systemic view need to be created and shared. Digital resources such as simulations that allow users to move from global to local scales, showing the connections between different phenomena, are particularly useful. The creation of shared databases of top-down teaching materials, validated by experts and enriched by teachers' experience, could also significantly accelerate the spread of the new approach.

Cultural resistance

The most difficult challenge is cultural resistance. Many people believe that the bottom-up approach, with its initial focus on details and classifications, is inherently 'more rigorous' and therefore more scientifically valid. This belief, rooted both among some teachers and in the collective imagination of what 'true science' is, can hinder the adoption of alternative approaches.

To overcome this resistance, it is essential to develop effective communication strategies that demonstrate, through empirical evidence, how the top-down approach not only does not sacrifice scientific rigor but enhances it by placing it in a meaningful context. Especially persuasive can be the involvement of professionals and researchers in the field who can testify how, in contemporary professional practice in geosciences, global understanding and the ability to navigate between different scales are core skills, often more relevant than encyclopaedic knowledge of detailed classifications.

Finally, documenting and sharing success stories in the implementation of the top-down approach, with a focus on outcomes in terms of deep awareness, student motivation, and the development of cross-curricular skills, is an effective strategy for gradually building consensus around the new teaching paradigm.

The following examples illustrate how the top-down approach can be implemented in practice. While the specific case studies mentioned can be adapted to local geological contexts—teachers should select examples relevant to their students' geographical location—the pedagogical structure remains universally applicable. The same framework can be used to study volcanoes in the Pacific Ring of Fire, the Mediterranean, or the East African Rift; river systems in any continental context; or mountain building processes in the Alps, Andes, Himalayas, or Appalachians.

TEACHING UNIT: "Understanding volcanoes"

Phase 1: The global context (1-2 lessons)

Start by visualizing the global distribution of volcanoes on a world map
Overlay the plate boundary map and discuss obvious correlations
Introduce the concept of plate tectonics as the "engine" of volcanism
Explore the different types of plate boundaries and their relationship to different types of volcanic activity

Phase 2: From global processes to regional phenomena (1-2 lessons)

Analyse how magma forms in different tectonic settings
Explore the differences between mid-ocean ridge, subduction zone, and hot spot volcanism
Relate these differences to the regional characteristics of different volcanic complexes
Examine representative case studies: Hawaii (hot spot), Andes (convergent margin), Iceland (oceanic ridge)

Phase 3: From region to local phenomenon (1-2 lessons)

Study the structure of a single volcanic edifice as a local manifestation of global processes
Analyse eruption products in relation to the tectonic context
Examine how the type of magma (basaltic, andesitic, rhyolitic) is determined by the tectonic position

Phase 4: From the phenomenon to the components (1-2 lessons)

Only now, study volcanic minerals and rocks as "witnesses" of the processes already seen
Explicitly link mineralogical composition and textures to the magma formation processes discussed in the previous phases

Assessment: Ask students to analyse a volcano not studied in class, placing it in the appropriate tectonic context and predicting characteristics such as eruption type, lava composition, and associated hazards, based on the global framework acquired.

TEACHING UNIT: "The hydrogeological system"

Phase 1: The global context (1-2 lessons)

Start with the planetary hydrological cycle and the global distribution of water resources
Illustrate the interconnection between surface water, groundwater, and atmospheric water as a unified system
Explore the interactions between the water cycle and large global climate systems
Discover how water distribution depends on plate tectonics (formation of basins, mountain ranges, etc.)

Phase 2: From global processes to regional systems (2-3 lessons)

Analyse how large river basins form in relation to regional tectonic settings
Study the differences between river systems in different tectonic regions (e.g., rivers in active margin vs. passive margin areas)
Explore how regional aquifers develop in different geological contexts
Examine representative case studies: the Mississippi (passive margin basin), the Ganges (foreland basin), the Rhine (post-orogenic basin)

Phase 3: From regional systems to local phenomena (2-3 lessons)

Study river forms (meanders, terraces, deltas) as local manifestations of regional processes
Analyse how lithology influences the characteristics of local aquifers
Examine erosion and transport phenomena in relation to regional drainage systems
Focus on the interaction between surface water and groundwater at the local scale

Phase 4: From phenomena to components (1-2 lessons)

Only at this point, examine river sediments and sedimentary rocks as "archives" of hydrogeological history

Study clay minerals and other components of sedimentary rocks as indicators of hydrological processes

Link rock porosity and permeability to the aquifer systems studied previously

Assessment: Ask students to analyse a local watershed, interpreting it as a part of a larger system. Students should be able to predict sediment characteristics, hydrological behaviour, and potential natural hazards based on their understanding of the global and regional framework.

The eruption of Erta Ale, Ethiopia, Danakil

Further reading

King, C., & Kennett, P. (2010). *Earth Science Investigations: 50 Earth Science Activities for Use in the Classroom and Outdoors*. Association for Science Education.

Occhipinti, S. (2013). *Earth Science Laboratory Experiments and Activities: Journey to the Centre of the Earth*. Department of Education and Culture, Autonomous Region of Valle d'Aosta. https://centrorisorsescienze.scuole.vda.it/images/Viaggio.pdf

·https://distancelearning.institute/curriculum-development/curriculum-development-models-top-down-vs-bottom-up/

CHAPTER 6.
GEOSCIENCES BEYOND ROCKS: MODELS AND SIMULATIONS

After exploring the top-down approach in the geoscience curricula, it is necessary to consider the teaching tools that make this transformation possible. As already illustrated, geosciences have a specific characteristic: they study processes that are difficult to observe directly or impossible to reproduce in the laboratory.

This specificity requires special teaching tools that allow us to visualize the invisible, compress geological time, and represent complex systems. Models therefore emerge not as simple teaching aids, but as true epistemological tools, essential for building a deep comprehension of geological phenomena.

This chapter will analyse how models, principally physical ones based on a hands-on approach, can make otherwise abstract concepts accessible, transforming the teaching of geosciences from a seemingly static and descriptive discipline into an adventure of exploration and discovery of the dynamic processes that shape our planet.

6.1 When a model becomes necessary

The introduction of models in geoscience teaching is particularly valuable in specific teaching situations, each characterized by obstacles that are unique to the direct perception of phenomena:

- Physical **inaccessibility of phenomena**: The study of the Earth's internal structure necessarily requires artifacts that represent layers and dynamics that cannot be observed directly. From the crust to the depths of the inner core, our knowledge derives from interpretive models based on indirect data such as seismic waves. Three-dimensional physical ones or digital simulations allow students to "see" what would otherwise remain confined to the imagination.
- **Incompatibility of time scales**: Geological processes such as orogenesis, sedimentation, or Wilson cycles unfold over tens or hundreds of millions

of Years. These tools allow enormous time intervals to be compressed into a few minutes or hours, making dynamics that would otherwise be abstract perceptible. A simulation of river delta formation can show processes that would take centuries to fully manifest in just a few minutes.

- **Complexity of natural systems**: Phenomena such as weather systems or biogeochemical cycles involve countless interdependent variables, making it virtually impossible to reproduce them in their entirety. Simplified representations allow us to isolate and visualize key components, facilitating knowledge of fundamental mechanisms without getting lost in the complexity of the complete system.
- **Limitations in experimental reproduction**: Even when it is technically possible to reproduce a geological phenomenon in the laboratory, reproduction is often only partial or on a small scale. Educational tools allow these limitations to be overcome, providing representations that, although simplified, capture the essence of the processes studied.
- **Changes in scientific understanding**: When the need arises to abandon established theories in favour of new interpretations that are more consistent with available data, visual representations facilitate this conceptual transition. The shift from a geocentric view of the Earth to the theory of continental drift, and subsequently to plate tectonics, was supported by visualizations that allowed new interpretations to be tested and communicated.

In all these situations, such pedagogical tools are not mere teaching aids but become essential instruments that connect natural phenomena to their conceptual awareness, bridging the gap between the observable and the imaginable.

The most significant contribution is the visualization of invisible or inaccessible processes: processes that take place underground, such as the circulation of fluids in aquifers or the generation of magma in subduction zones, can be explored through transparent materials or exposed sections. This approach allows us to construct adequate mental representations of otherwise abstract phenomena.

Control of time scales is another crucial advantage. Educational simulations allow processes that would take millennia to be compressed into a few minutes or hours of observation. A river erosion demonstration can show in real time what would take centuries to fully manifest in nature.

Managing complexity through controlled simplification makes possible specific aspects or mechanisms of inherently complex geological systems to be isolated.

An atmospheric circulation model can initially focus only on the effects of Earth's rotation, then gradually incorporate other factors such as temperature differences or orography.

6.2 The hands-on approach: a conscious choice

In the contemporary educational landscape, characterized by a growing availability of digital and virtual resources, the choice to favour a hands-on approach in the teaching of geosciences is a conscious methodological decision based on solid pedagogical foundations. Although there are numerous high-quality simulation software and interactive applications, direct experience with tangible physical representations continues to demonstrate effectiveness in school contexts of all levels.

Active interaction with teaching materials stimulates an experimental approach, which is essential in scientific education. By manipulating variables, altering initial conditions, and observing how these changes affect the evolution of the system represented, students experience the scientific process firsthand. This direct experimentation promotes discovery learning and develops fundamental methodological skills based on the formulation and testing of hypotheses.

Physical experience with educational tools activates cognitive processes, involving different sensory channels and creating richer and more memorable educational experiences. When manipulating layers of sand to simulate tectonic deformations, observing the formation of waves in a water tank, or modelling the relief of a territory with different materials, students activate multiple connections that facilitate understanding and memorization. This manipulative dimension of learning is especially valuable in geosciences, where many central concepts refer to concrete physical and material processes.

The experience of educators confirms the effectiveness of this approach. Many teachers, faced with the detached gaze of students during theoretical lessons, have discovered how even rudimentary constructs built with simple materials can radically transform the dynamics of learning. Interdisciplinary collaboration, involving teachers from different subjects and the students themselves in the construction of teaching aids, further enriches the learning experience.

Observations show that concepts learned through interaction with physical representations tend to remain more easily in the memory. Even years later, former students recall these experiences with surprising accuracy, often citing them as key moments in their educational journey. It is not necessarily sophisticated equipment that determines teaching effectiveness, but rather the demonstration's ability to capture the essence of a complex process, making it understandable while maintaining its scientific integrity.

The construction of these tools themselves is a key component of the educational experience. When students actively participate in creating a volcanic replica, setting up an erosion experiment, or assembling a seismic wave simulator, they engage in a process that requires preliminary knowledge, planning, and problem solving. This direct involvement promotes conceptual appropriation and the development of "manual intelligence," an increasingly valuable skill in an age of growing virtualization.

UNESCO, in its document "Hands-on Science," emphasized that "the best way to increase students' interest in science is to use teaching methods that appeal to the curiosity and creativity that characterize all children." The hands-on approach stimulates motivation, a key element for effective learning. Direct experience with manipulatives awakens natural curiosity: by touching and experimenting, students spontaneously begin to ask questions and formulate hypotheses. The hands-on approach is based on a long pedagogical tradition that, from Aristotle to John Dewey, has recognised the value of direct experience. The principle of "learning by doing" works particularly well in geosciences: it is more natural to start from concrete experience and move on to concepts than to start from abstract theory.

Hands-on activities also promote the development of cross-cutting skills such as teamwork, critical thinking, and problem-solving. When groups of students collaborate in the analysis of a geological representation, cooperative learning dynamics are created that reproduce the processes of social construction of scientific knowledge.

The effectiveness of this approach requires continuous updating of materials and methods. Teaching instruments must grow alongside new scientific discoveries, adapting to different contexts thanks to collaboration between researchers, experts, and teachers in the field.

Despite the potential of digital technologies, the hands-on approach remains central to the teaching of geosciences, offering concrete experiences of the phenomena studied. It is not a question of rejecting digital technology, but of recognising the irreplaceable value of direct experience with tangible materials, especially in a discipline where touching and manipulating are fundamental. As will be explored in a dedicated chapter, modern simulation technologies are complementary tools that, when integrated with a hands-on approach, expand teaching possibilities without replacing direct experience.

6.3 Types of models for teaching geosciences

Modelling in geosciences can take many forms, each with specific characteristics and educational potential.

Physical models use common materials to simulate complex geological processes. For example, layers of coloured sand can represent sedimentary sequences deformed by tectonic movements; gelatine can simulate the elastic behaviour of the Earth's crust; mixtures of glycerine and water can reproduce convective cells in the mantle. These representations allow tangible interaction and direct observation of the simulated processes.

Conceptual frameworks represent abstract relationships between components of a geological system. The hydrological cycle, for example, identifies the main water reservoirs (oceans, atmosphere, glaciers, groundwater) and the flows between them. These constructs, often represented through flow diagrams or conceptual maps, are particularly useful for visualizing interconnections and feedback in complex systems.

Three-dimensional visualizations, created with digital technologies or physical materials, help to understand complex geological structures such as folds, faults, or volcanic systems. 3D representation is especially important in geosciences, where understanding spatial relationships is often crucial for interpreting processes. Augmented reality technologies have further enriched these possibilities, allowing geophysical and geological data to be superimposed on real topographic models.

Mathematical and computational approaches translate geological processes into equations and algorithms. Although more abstract, they allow for precise simulations and the exploration of multiple scenarios, incorporating numerous variables and representing evolutions over long time intervals.

6.4 Characteristics of an effective model

The educational effectiveness of a model does not depend on its technical complexity but on its ability to illuminate core concepts. An educationally valid model should meet certain essential criteria that determine its pedagogical and scientific value.

Scientific accuracy: even if simplified, a model must maintain the necessary relationships between the elements of the system, avoiding errors that could create false ideas. A volcano model can be simple but must correctly show how magma rises and the differences between types of eruptions.

Essentiality: a good model focuses on the main aspects of the phenomenon, eliminating unnecessary details that could cause confusion. This 'economy' helps students to focus on the fundamental mechanisms without getting lost in the details.

Accessibility: to be truly useful, a model must be able to be made with common materials and simple procedures. This makes it usable even where resources are limited, democratizing access to quality educational experiences.

Interactivity: an effective model allows students to act on it, modifying parameters and observing the results. This interaction transforms the experience from passive to active, involving students in a process of discovery.

Ability to stimulate: a truly effective model does not just provide ready-made answers but generates questions and stimulates curiosity. When students spontaneously begin to ask questions and make hypotheses, the model becomes a tool for scientific research.

6.5 Preliminary analysis of the target audience

An effective representation must be tailored to the cognitive abilities and prior knowledge of the students. The same teaching tools can work for different levels of education, but the approach and level of detail must be adapted, as well as any connections to other disciplines.

A river erosion demonstration for primary school will focus on visible and concrete aspects, while the same phenomenon for university students may include quantitative variables such as flow velocity and sediment grain size. Educational materials must build cognitive bridges between what students already know and the new concepts to be introduced.

The pedagogical tool must integrate harmoniously into the broader curriculum, linking up with other learning experiences and contributing to a coherent view of the subject. A volcano simulation, for example, acquires greater meaning when included in a course that covers the study of rock types, plate tectonics, and geological risk.

Teacher Involvement

To ensure the effectiveness of educational resources, teacher involvement is essential. Teachers should not be mere users but active partners in design, bringing their field experience and knowledge of their students' needs. This collaboration significantly increases the practical applicability of the materials.

Equally important is curricular integration. Hands-on activities should not be isolated but integrated into the overall curriculum, supporting the course's educational objectives. A demonstration loses effectiveness if it is not embedded in a structured learning pathway.

The final goal is to develop professional autonomy: the teacher must not only use these tools confidently but creatively adapt them to the specific needs of their own context, personalizing and improving them.

The adoption of innovative approaches often requires a rethinking of the educational method. Teachers must be supported in this transition which implies a change in their role: from transmitters of knowledge to facilitators of active learning experiences.

The use of models in geoscience teaching presents numerous established benefits. Modelling offers significant advantages, starting with the visualization of invisible processes. Physical representations make perceptible phenomena that are otherwise inaccessible, such as tectonic plate movements or mantle circulation, helping to understand complex and abstract concepts.

They also allow for the simplification of complex geological systems, isolating specific aspects to create functional demonstrations. Students can thus understand fundamental principles without getting lost in the numerous variables of natural systems.

Particularly important is the possibility of active experimentation. Students can test hypotheses, modify variables, and observe effects in real time. Physical tools offer a multisensory experience - seeing, touching, manipulating - that creates richer and more memorable learning.

An additional advantage is the possibility of creating effective teaching aids with common and inexpensive materials, making quality teaching accessible even where resources are limited.

Educators must also recognise the intrinsic limitations of this approach. The main risk is excessive simplification: reducing complexity too much can lead to incomplete appreciation, losing sight of the intricate interactions of natural systems.

Another challenge concerns temporal and spatial scales. Representations must compress processes of millions of years into manageable times and reduce immense dimensions to manipulable scales, risking altering the perception of the real dimensions of geological phenomena.

It is also difficult to show how natural processes occur simultaneously: often one ends up isolating single mechanisms instead of representing them together. Furthermore, materials can only approximate the real properties of rocks and fluids.

Finally, contextualization of the experience is always needed. Without connecting modelling to direct observation, students risk mistaking the representation for reality instead of for a simplified version of it.

Toward an integrated approach

To maximize the benefits of modelling while minimizing its limitations, it is essential to adopt an integrated approach. Educational tools acquire full pedagogical meaning when they are embedded in a coherent educational pathway that progressively develops comprehension of geological phenomena, creating conceptual connections between different learning experiences.

The support of complementary methodologies is central: hands-on demonstrations should be supported by other educational strategies, such as direct field observation, study of real cases, and use of digital technologies, each with their own specific strengths.

The maximum effectiveness is achieved when students are guided through metacognitive reflection on the representations themselves, recognising both their potential and limitations. This metacognitive awareness transforms the experience from simple practical activity to authentic scientific learning opportunity.

6.6 Models as tools for geoscientific literacy

The conscious use of models in geoscience teaching transcends simple content transmission: it contributes to the development of true "geoscientific literacy," the ability to interpret the world through the perspective of geosciences. This competence integrates understanding of deep temporal scales, perception of systemic interconnections, three-dimensional spatial thinking, and awareness of interactions between natural processes and human activities.

In an era of unprecedented environmental challenges, this literacy becomes essential not only for specialists, but for all citizens. Pedagogical tools, by making the invisible visible and compressing geological time into experiences, transform passive observers into conscious interpreters of the environment surrounding them.

Educational Models: simplicity and effectiveness for understanding complexity

The teaching aids that follow are characterized by their constructive simplicity and contained costs, often achievable with common-use materials easily obtainable in any school. Some of these demonstrations are now established classics in geoscience education, others less known but equally valuable, still others might even appear trivial in their elementary nature.

But precisely this apparent simplicity constitutes their strength: the proven effectiveness in making complex geological phenomena comprehensible, in visualizing temporal sequences difficult to imagine, and in facilitating awareness of processes developing on spatial and temporal scales often inaccessible to direct experience. These educational tools perfectly embody the hands-on approach previously illustrated, transforming abstract concepts into concrete and manipulable experiences. The visual approach that characterizes them, even when representations may seem somewhat naïve or simplified compared to the reality of phenomena, proves fundamental for building that first intuitive knowledge that constitutes the basis for more sophisticated learning.

Their true richness lies in their ability to make geoscience teaching universally accessible, regardless of the school's technological equipment. They thus allow every student to "touch with their hands" the great processes that have shaped and continue to shape our planet.

The examples presented here include both widely adopted classics of geoscience education and others developed through direct teaching experience in specific geographical contexts. While some reference particular geological settings—such as Mediterranean tectonics or Alpine structures—the underlying pedagogical principles remain universally applicable. Teachers are encouraged to adapt these demonstrations to their own regional geology, substituting local examples that will resonate more meaningfully with their students while maintaining the same educational objectives and structural approaches.

CONTINENTAL DRIFT MODEL
DESCRIPTION AND STRUCTURE: Puzzle made with continents cut out of different materials (stiff cardboard, plywood, wood) according to their current coastlines. Continental bridges and minor continental fragments can be represented to complete paleogeographic reconstructions.
EDUCATIONAL PURPOSE: The model demonstrates how the continents fit together perfectly in the reconstruction of Pangea, providing tangible evidence for Wegener's theory of continental drift. It allows the continuity of mountain ranges, geological formations, and glacial deposits to be visualized. The model develops spatial and temporal reasoning skills on geological scales, preparing students to understand the physical mechanisms of plate tectonics and introducing the concept of deep time in Earth's history.
EXPERIMENTAL ACTIVITIES: It is possible to reconstruct Pangea by observing the geometric correspondences of coastal forms but also the alignment of mountain chains and the distribution of characteristic fossil fauna and flora. The Atlantic coasts of Africa and South America can be compared in detail to highlight the correspondences of continental margins, the geological correspondences between Brazil and Ghana, Appalachians and Caledonidi, the distribution of Mesosaurus between South America and southern Africa, and of Permian tillites.

PANGEA MODEL
DESCRIPTION AND STRUCTURE: Puzzle made of rigid cardboard or plywood panels that reproduce the current continents according to their precise shapes, shaped following the margins of the continental platforms rather than simple coastlines. The model could use coloured markers to highlight mountain ranges, glacial deposits, and fossiliferous areas that align in the reconstruction of the supercontinent.
EDUCATIONAL PURPOSE: The model enables to visualize how the continents fit together perfectly in the configuration of Pangea. It also develops spatial reasoning and three-dimensional visualization skills.
EXPERIMENTAL ACTIVITIES: Guided reconstructions of Pangaea can be carried out starting from the current continents, comparing in particular the Atlantic coasts of Africa and South America to highlight geometric correspondences. The mountain ranges that line up in the Pangaea configuration can be located and the distribution patterns of identical fossils on continents that are now separated can be traced. Simulation of the progressive fragmentation of the supercontinent support discussion of the geological, paleontological, and paleoclimatic evidence supporting the theory, linking concrete observations to processes on a planetary scale.

MODEL OF THE OPENING OF THE ATLANTIC OCEAN

DESCRIPTION AND STRUCTURE: Puzzle made of rigid cardboard or plywood panels reproducing the African, European, and American continents and the position of present-day Italy, while coloured strips of different shades represent bands of oceanic crust of progressively increasing age

EDUCATIONAL PURPOSE: The model visualizes the process of ocean floor expansion. It directly links the opening of the Atlantic Ocean to the progressive separation of Pangaea. It prepares students to understand the Wilson cycle and the processes of oceanic crust formation and destruction

EXPERIMENTAL ACTIVITIES: It is possible to simulate the progressive expansion of the Atlantic Ocean through different geological eras. The formation of the symmetry of crustal bands of equal age on both sides of the ridge can be reconstructed, and the age of the oceanic crust can be calculated as a function of the distance from the centre of expansion

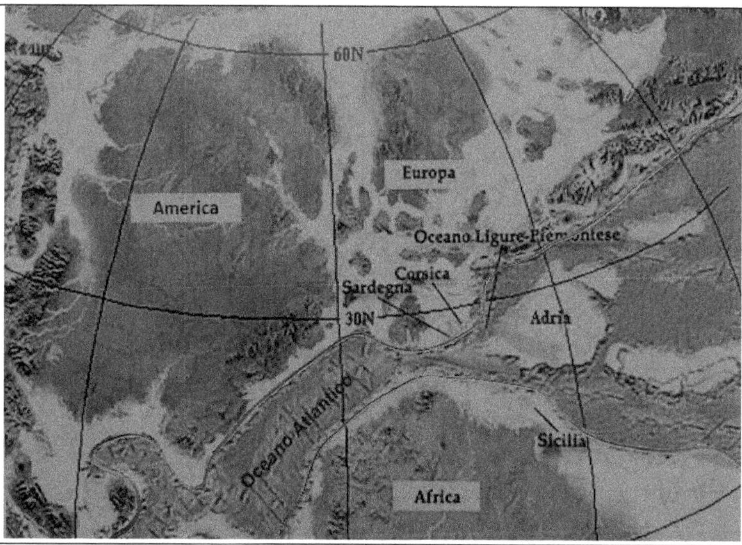

MODEL OF THE APULIAN FLACK AND THE FAULT SYSTEM OF THE MEDITERRANEAN AREA

DESCRIPTION AND STRUCTURE: Plywood model reproducing the Mediterranean basin with mobile plates of the African plate, the Apulian microplate (Adriatic) and the Eurasian plate. Coloured arrows indicate the directions of movement, while dotted lines highlight the main active fault systems.

EDUCATIONAL PURPOSE: The model provides an understanding of the complex geometry of the Mediterranean tectonic system and the role of the Apulian microplate in the formation of the Italian mountain ranges. It shows how current tectonic movements are directly linked to seismic activity in Italy and explains the formation of the Alps and Apennines through different but interconnected tectonic processes.

EXPERIMENTAL ACTIVITIES: It is possible to simulate the convergence between the Apulian plate and the European margin that led to the formation of the Alps, as well as to reconstruct the evolution of the Apennines through the subduction processes of the Ionian crust. The main active faults can be identified and correlated with the distribution of historical seismicity in the Red Sea.

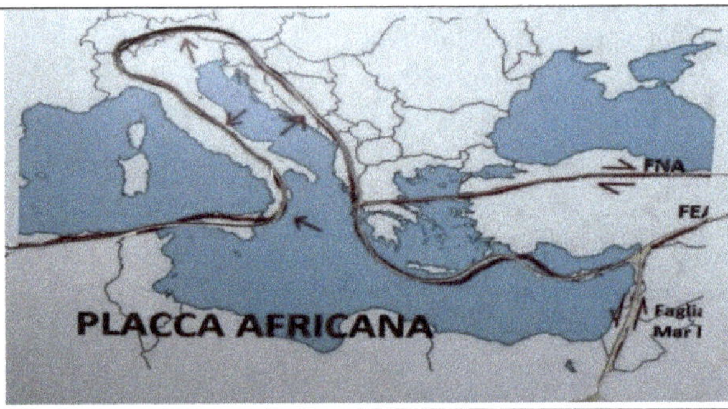

This model, developed for the specific context of Italian tectonics, exemplifies how regional geological complexity can be made comprehensible through physical modelling. Teachers in other regions can apply the same pedagogical approach to their own tectonic settings: the interaction between the Pacific, North American, and Juan de Fuca plates on the west coast of North America; the complex collision zone between the Indian and Eurasian plates forming the Himalayas; or the Caribbean plate interactions with North and South American plates. The educational value lies not in the specific geographical example but in understanding how microplate dynamics influence regional geology and seismic hazards.

PLATE TECTONICS MODEL
DESCRIPTION AND STRUCTURE: Panel made of different materials (rigid cardboard, plywood, or wood) with sections representing the main tectonic plates of the Earth, shaped according to their actual shapes and dimensions on a reduced scale. Each plate shows the simulation of the three basic types of margins: divergent, convergent, and transform
EDUCATIONAL PURPOSE. The model shows how the relative movements of tectonic plates generate earthquakes, volcanic activity, and mountain ranges in specific locations on the planet, demonstrating the direct relationship between deep geodynamic processes and observable surface phenomena. It facilitates comprehension of the different types of plate margins and associated processes, linking plate tectonic theory with the global distribution of geological hazards. It develops the ability to read and interpret global tectonic and seismic maps
EXPERIMENTAL ACTIVITIES: It is possible to move plates according to their actual movements and predict where ocean ridges, subduction trenches, and mountain ranges will form, verifying the predictions with the actual distribution of earthquakes and volcanoes. Continental collisions, ocean basin openings, and subduction processes can be simulated, observing how these influence the Earth's morphology. The comparison between the theoretical model and the actual distribution of geological phenomena facilitates the discussion of the root causes of natural hazards and their predictability

MODEL OF THE EARTH'S INTERIOR	
DESCRIPTION AND STRUCTURE: A sectioned model that reproduces the internal structure of the planet with different colours to represent, on the one hand, the physical characteristics, lithosphere and asthenosphere, liquid outer core and solid inner core and, on the other hand, the chemical composition, crust, mantle, and core. The model highlights the main discontinuities (Moho, Gutenberg).	
EDUCATIONAL PURPOSE: The model makes possible to visualize the layered structure of the Earth and understand how variations in the composition, density, and physical state of materials influence geodynamic processes. It facilitates understanding of the origin of the Earth's magnetic field, the convection mechanisms of the mantle, and the propagation of seismic waves. It links deep physical properties with observable surface phenomena, from plate tectonics to volcanism.	
EXPERIMENTAL ACTIVITIES: It is possible to observe the different interfaces and discuss the main seismic discontinuities. The densities and compositions of the different layers can be compared, and how variations in physical properties influence wave arrival times can be discussed. Considerations on convective cells in the mantle allow deep processes to be linked to surface plate movements.	

ISOSTASY MODEL

DESCRIPTION AND STRUCTURE: A system of wooden or plastic blocks filled with lead weights of different sizes and densities floating in a transparent container filled with spongy material to simulate the behaviour of the Earth's crust on the asthenospherc mantle. The blocks represent different types of crust (thin and dense oceanic, thick and light continental) and mountain masses of varying thickness. A system of removable weights simulates different loads.

EDUCATIONAL PURPOSE: The model demonstrates the principle of isostatic equilibrium and explains why mountain ranges deep roots have while ocean basins correspond to thin crust. It illustrates the mechanisms of post-glacial uplift, sediment load subsidence, and vertical adjustments of the Earth's crust. It facilitates perception of relative sea level changes and the processes of marine and lake terrace formation.

EXPERIMENTAL ACTIVITIES: Loads can be added and removed to observe the resulting vertical adjustments and measure the response times of the system. The formation and melting of ice caps can be simulated by observing the effects on the relative sea level, as well as the sedimentary loading of subsiding deltas and basins. The comparison between oceanic and continental crust allows the altimetric differences between continents and ocean floors to be explained.

TECTONIC DEFORMATION MODEL – SAND BOX – VARIOUS MODELS

DESCRIPTION AND STRUCTURE: System consisting of transparent containers filled with alternating layers of granular materials of different colours (fine sand, flour, ground coffee) that reproduce stratigraphic sequences. A side piston, which can be operated manually or mechanically, applies controlled compressive or tensile forces.

EDUCATIONAL PURPOSE: The model visualizes deformation processes that would take millions of years in nature, allowing the formation of folds, faults, overthrusts, and complex structures to be observed in real time. It links deep tectonic forces with geological structures observable on the ground and facilitates the interpretation of geological sections and structural maps.

EXPERIMENTAL ACTIVITIES: Different stress regimes can be applied to generate structures comparable to those observed in real mountain ranges, varying the speed and intensity of the forces to observe brittle or ductile behaviour. Analysis of the resulting geometries enables the deformation history to be interpreted and correlated with regional geodynamic evolution.

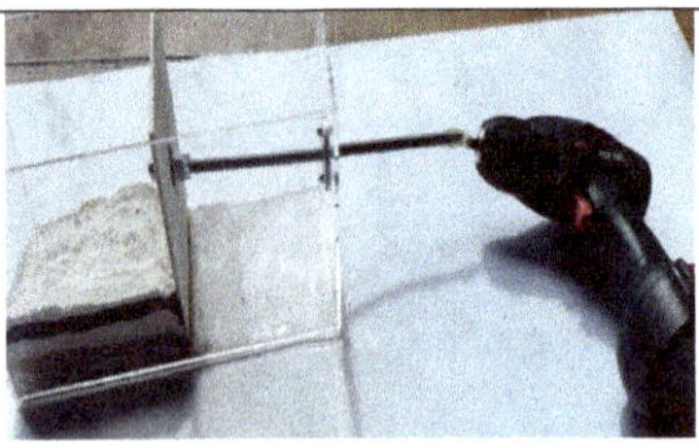

OCEAN CIRCULATION MODEL
DESCRIPTION AND STRUCTURE: Container divided into sections, or feedable at different levels, allowing the creation of water masses at different temperatures, salinity, and density to simulate ocean stratification and thermohaline circulation mechanisms. Variable speed pumps and dye systems allow the visualization of surface and deep currents
EDUCATIONAL PURPOSE: The model illustrates the physical mechanisms that drive global ocean circulation, linking local processes (density, convection) with complex planetary phenomena such as thermohaline circulation and atmosphere-ocean energy exchanges. It facilitates perception of the role of oceans in global climate regulation and planetary heat distribution. It demonstrates the interconnection between physical, chemical, and climatic processes in Earth systems.
EXPERIMENTAL ACTIVITIES: Different density configurations can be created to observe the formation of deep and surface currents, adding dyes to visualize the paths of water masses and mixing times. Phenomena such as the formation of deep polar waters and upwelling can be simulated. Altering the initial conditions allows the effects of climate change on global ocean circulation to be discussed.
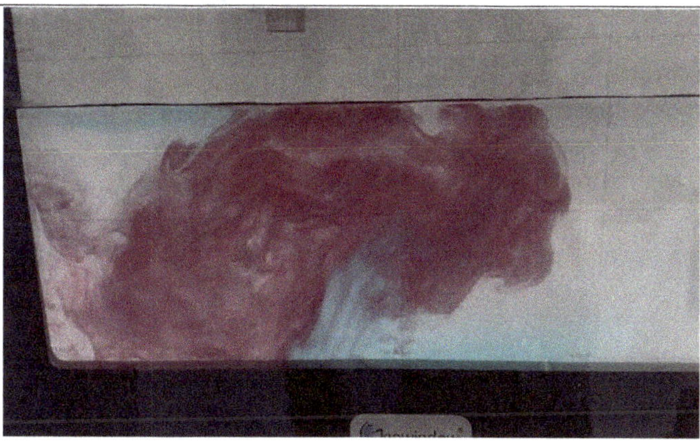 Created with Geowindow - https://www.geowindow.de

VOLCANO AND EARTHQUAKE DISTRIBUTION MODELS

DESCRIPTION AND STRUCTURE: Interactive planisphere with LED system representing active volcanoes and seismic epicentres. The base of the model reproduces the planisphere, while transparent layers allow different information to be superimposed: distribution of seismic activity, location of volcanoes by type of activity, correlations with deep tectonic structures.

EDUCATIONAL PURPOSE: Visually demonstrates the correlation between global distribution of earthquakes, volcanoes, and plate margins. Highlights how geological phenomena follow precise patterns linked to Plate Tectonics.

EXPERIMENTAL ACTIVITIES: Students activate phenomena to observe spatial correlations and compare the observed distribution with that predicted by Plate Tectonics, the relationship between different types of seismic and volcanic activity and geodynamic contexts. They analyse zones of particular seismic concentration (Pacific Ring of Fire, Alpine-Himalayan system) and discuss implications for exposed populations.

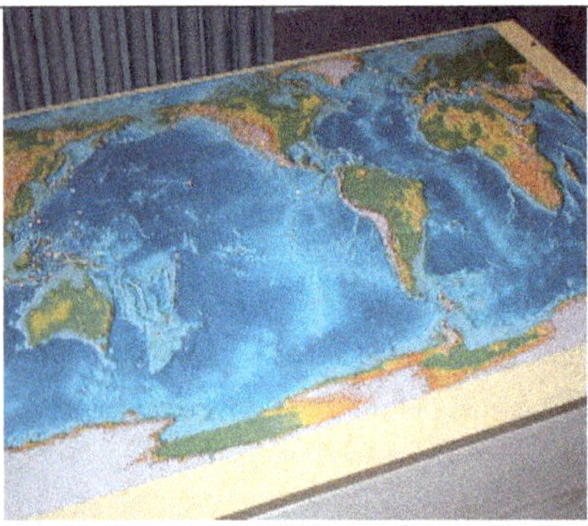

EARTHQUAKE SIMULATION MODELS

DESCRIPTION AND STRUCTURE: Vibrating table equipped with scale models of buildings made of different materials (wood, cardboard, plasticine) that reproduce different types of construction and degrees of structural flexibility. Sensors record the vibrations transmitted to the structures, while a wave generator produces artificial earthquakes with different frequency and duration characteristics.

EDUCATIONAL PURPOSE: The model allows students to understand how seismic waves propagate through different materials and how the response of buildings depends on the dynamic characteristics of both the earthquake and the structures. It illustrates the principles of earthquake engineering and the importance of appropriate design in seismic zones.

EXPERIMENTAL ACTIVITIES: Students generate artificial earthquakes with different characteristics and quantitatively observe damage on structures of various types, testing the effectiveness of different seismic systems. They develop scientific awareness of seismic risk and mitigation strategies through technology.

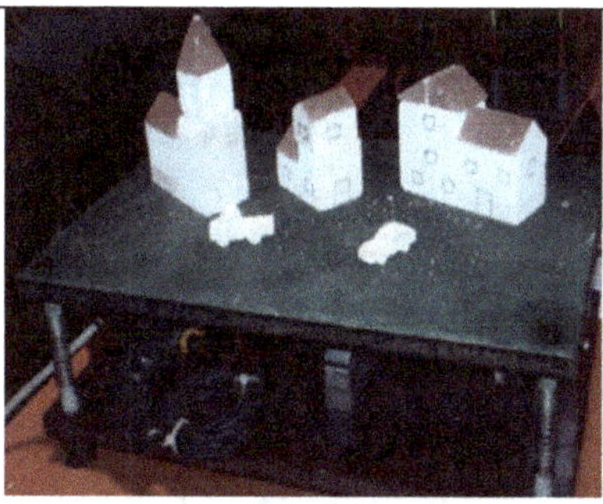

VOLCANO MODEL

DESCRIPTION AND STRUCTURE: A sectional model showing the complete internal structure of a volcanic system, from the deep magma chamber to the feeding conduits and the surface volcanic edifice. Different types of magma are simulated with materials of varying viscosity to reproduce distinct eruptive behaviours. The model includes systems to simulate effusive, explosive, and mixed eruptions to observe the effects on the eruptive mode

EDUCATIONAL PURPOSE: The model focuses attention on the key relationship between magma chemistry, eruption type, and volcanic building shape, going beyond a purely spectacular approach to develop scientific understanding of the processes. It links chemistry, physics, and geomorphology in the explanation of volcanic phenomena.

EXPERIMENTAL ACTIVITIES: Materials with different viscosities can be used to simulate different magmas, observing how composition influences eruptive behaviour. Effusive eruptions that produce lava flows can be compared with pyroclastic explosions that generate fiery clouds and fallout deposits. Analysis of the random location of the eruption facilitates considerations on risk and planning for prevention purposes

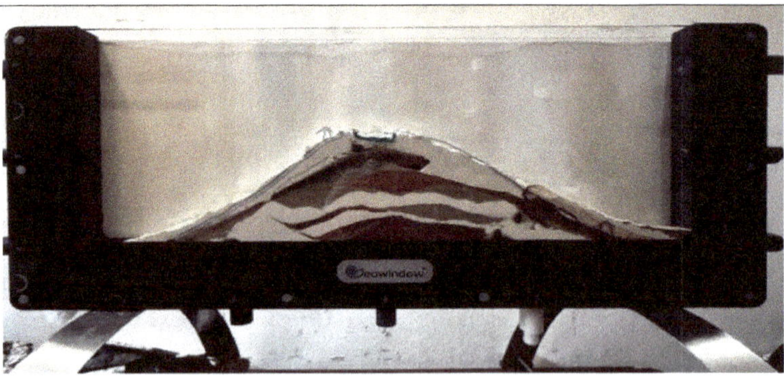

Made with Geowindow - https://www.geowindow.de

RIVER EROSION MODEL
DESCRIPTION AND STRUCTURE: Tilting tray filled with sediments of different grain sizes. Controlled irrigation system simulates precipitation of varying intensity, while drainage channels reproduce the hydrographic network.
EDUCATIONAL PURPOSE: Allows real-time observation of channel, meander, and alluvial fan formation. Demonstrates how slope, discharge, and grain size influence erosive and depositional processes.
EXPERIMENTAL ACTIVITIES: Students modify one parameter at a time and verify predictions through controlled experiments. They simulate flood events, anthropogenic modifications of watercourses, and effects of land use on runoff.

HYDROGRAPHIC SYSTEM MODEL
A system that reproduces a complete hydrographic basin from its source to its mouth, created using different topographic surfaces that represent a hydrographic network. The structure includes headwater reservoirs to simulate different supply conditions (springs, lakes, glaciers), variable-section channels to represent the morphological evolution of watercourses from high mountains to plains, and flow control systems to simulate seasonal hydrological regimes. The model integrates anthropogenic elements such as diversions, dams, regulation works, and urbanization of the basin.
EDUCATIONAL PURPOSE: Facilitates comprehension of the watershed concept and the relationships between morphology, climate, vegetation, and surface hydrology. It is possible to visualize the drainage network formation processes by connecting mountain erosive processes, riverbed transport, and sedimentation in plain areas, highlighting how local modifications can propagate through the entire hydrographic system.
EXPERIMENTAL ACTIVITIES: Students simulate different hydrological regimes by observing the propagation of flow variations, analysing concentration times. They simulate flood events by observing overflow dynamics and identifying critical areas for safety. They can test effects of urbanization through progressive impermeabilization, observing modifications in response times and flood peaks.

SOIL PERMEABILITY MODEL	
DESCRIPTION AND STRUCTURE: System of transparent containers filled with different types of materials (clay, silt, fine sand, coarse sand, gravel) to reproduce soils with increasing permeability. A system of dyes allows the preferential paths of water through the different materials to be visualized.	
EDUCATIONAL PURPOSE: The model illustrates the basic hydraulic properties of soils and their relationship with grain size, porosity, and structure. It facilitates knowledge of the processes of infiltration, water retention, and water movement. It links the physical properties of soils with applied issues such as irrigation, agricultural drainage, and water resource management.	
EXPERIMENTAL ACTIVITIES: It is possible to quantitatively compare the infiltration rates and retention capacities of different soil types by measuring percolation times and retained water volumes. Different rainfall intensities can be simulated to observe differential phenomena. The use of dyes shows preferential accumulation zones and underground flow paths.	
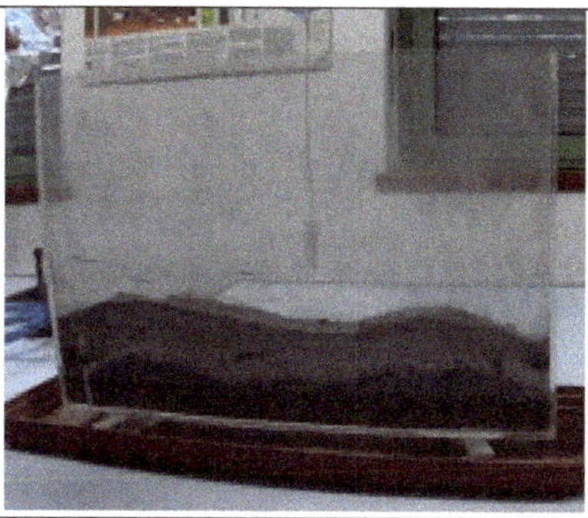	

RAIN SIMULATION MODEL

DESCRIPTION AND STRUCTURE: System consisting of a drip sprinkler positioned above an inclined surface on which different materials are placed (bare soil, vegetation, impermeable surface). The system allows the intensity, duration, and spatial distribution of simulated precipitation to be varied.

EDUCATIONAL PURPOSE: The model demonstrates the effects of rainfall intensity on erosion, infiltration, and runoff processes, highlighting the role of vegetation cover and impermeabilization in stormwater management. It facilitates understanding of the processes linking meteorology, paedology, and surface hydrology. It develops awareness of the environmental impacts of urbanization and agricultural practices.

EXPERIMENTAL ACTIVITIES: It is possible to quantitatively compare the effects of different rainfall intensities on surfaces with and without vegetation cover by measuring erosion rates. Extreme weather events can be simulated, and the effectiveness of different soil protection strategies can be evaluated.

GROUNDWATER MODEL
DESCRIPTION AND STRUCTURE: Transparent container layered with materials of different permeability (coarse sand, fine sand, gravel, impermeable clay lenses) that reproduce the structure of natural aquifers. The system includes piezometers to measure groundwater levels at different points and feeding systems to simulate recharge from precipitation. Different dyes or hot water allow the movement of water to be traced and pollution phenomena to be simulated.
EDUCATIONAL PURPOSE: The model visualizes the movement of groundwater and the formation of unconfined and confined aquifers, illustrating central concepts such as piezometric surface, saturated and unsaturated zones, and flow directions. It demonstrates the impact of human activity on groundwater resources through pumping, pollution, and changes in recharge. It facilitates awareness of the vulnerability of aquifers and protection strategies.
EXPERIMENTAL ACTIVITIES: It is possible to simulate recharge and discharge cycles of the aquifer by observing changes in piezometric levels and changes in flow directions. Coloured contaminants or hot water can be introduced and measured with thermometers in piezometric tubes to study the dispersion and attenuation mechanisms of pollutants.

RISK VALLEY MODEL
DESCRIPTION AND STRUCTURE Three-dimensional model reproducing a river valley characterized by multiple types of geological risk. The structure includes unstable slopes to simulate landslide or avalanche zones and floodplains to highlight flood-prone areas. The model is equipped with movable elements (houses, infrastructure, production activities).
EDUCATIONAL PURPOSE: Develops skills for integrated assessment of geological risks. Illustrates how apparently independent events can chain together in complex risk scenarios. Promotes perception of the key role of territorial planning in reducing vulnerability
EXPERIMENTAL ACTIVITIES: Students strategically position settlements and infrastructure by evaluating exposure to different risks. They analyse the effectiveness of alternative mitigation strategies and construct integrated risk maps.

GEOLOGICAL MODEL OF THE AOSTA VALLEY

DESCRIPTION AND STRUCTURE: Panel and geological section reproducing the complex structure of the Western Alps using the example of the Aosta Valley, showing overthrust tectonic units and intrusive plutons. The model includes removable elements that allow the geometric relationships between the different geological units to be observed, while a cross-section illustrates the deep structure of the Alpine chain. Standard colour codes identify the different geological formations.

EDUCATIONAL PURPOSE: The model helps to understand the structural complexity of mountain ranges and the geological history of the Alps through a concrete and geographically significant example. It illustrates concepts such as thrust faults, metamorphism, and plutonism in a real geological context. It facilitates the interpretation of complex geological maps and the development of geological landscape reading skills.

EXPERIMENTAL ACTIVITIES: It is possible to reconstruct the chronological sequence of tectonic, magmatic, and morphological events by observing the geometric and overlapping relationships between the different rock units. The history of deformation can be interpreted by analysing folds and thrust surfaces. The comparison between geology and current morphology allows for discussion of the Quaternary evolution of the Alpine chain and the effects of glacial modelling.

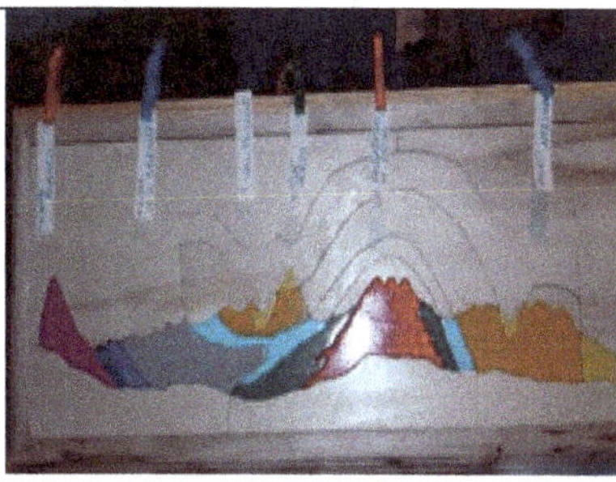

This model was developed and extensively tested in educational contexts in the Italian Alps. It serves as a detailed example of how complex mountain belt geology can be represented through physical modelling. While this specific model focuses on Alpine geology, the same approach can be adapted to any mountain belt worldwide—the Appalachians, Andes, Rockies, Atlas, or Urals—by modifying the geological units while maintaining the pedagogical structure that illustrates thrust tectonics, metamorphism, and plutonism.

The models presented in this chapter range from globally applicable classics to examples developed in specific regional contexts. What unites them is not geographical universality but pedagogical effectiveness: each demonstrates how complex geological concepts can be made tangible through thoughtful model design. Educators worldwide are encouraged to view these as templates for developing their own regionally appropriate examples, adapting geological content to local contexts while maintaining the hands-on, Inquiry-Based pedagogical approach that makes physical modelling so effective in geoscience education

The contact between the European and American plates - Thingvellir National Park - Iceland

Further reading

1. UNESCO (2010). *Hands-on Science: Promoting Scientific Inquiry in Education*. UNESCO Publishing.

2. Dewey, J. (1938). *Experience and Education*. Kappa Delta Pi. [Classic on the pedagogy of direct experience]

3. Gilbert, J.K. & Justi, R. (2016). Modelling-based Teaching in Science Education. Springer International Publishing. https://link.springer.com /chapter/10.1007/978-3-319-29039-3_4

4. Oreskes, N. (2007). "Models and the History of Science." Earth Science History, 26(1), https://history.ucsd.edu/_files/faculty/oreskes-naomi/roleo fmodels.pdf

5. McLaughlin, J. A., & Bailey, J. M. (2022). Students need more practice with spatial thinking in geoscience education: a systematic review of the literature. *Studies in Science Education*, *59*(2),204. https://doi.org /10.1080/03057267.2022.2029305

6. Reynolds, S.J. et al. (2005). "Visualization in Undergraduate Geology Courses." In *Visualization in Science Education* (pp. 253-266). Springer.

CHAPTER 7.
MINERALS AND ROCKS: STORYTELLERS

Geosciences often seem like a discipline far removed from everyday life, confined to museum display cases. In particular, traditional teaching about minerals and rocks is often reduced to sterile classifications. This chapter proposes a revolution: transforming "rocks" into protagonists of stories that connect the planet's past to today's challenges. Each "rock" contains fascinating stories that connect geology to everyday life, history, economics, and the great processes of the planet. The narrative approach presented here aims to awaken students' curiosity by revealing how "rocks" are eloquent witnesses to the geological, technological, and cultural transformations that have shaped—and continue to shape—our existence.

The examples presented in this chapter draw from diverse geographical and cultural contexts, with particular attention to Mediterranean history and Italian geological heritage. While some references are regionally specific, the pedagogical approach—transforming minerals and rocks from abstract classifications into compelling narratives—is universally applicable. Teachers are encouraged to complement these examples with stories from their own regional geology and cultural history, creating connections that resonate most powerfully with their students lived experiences.

Examples spanning archaeology, geopolitics, art, and digital technologies reveal, the possibilities of interdisciplinary teaching, which reveals the fabric that links minerals to human history.

7.1 Minerals as invisible protagonists of human history

A different approach to teaching minerals is to present them as elements that have shaped human history. Iron is an emblematic element: the transition from copper/bronze to iron (1200-1000 BC) redefined the balance of power in the ancient world. The first iron artifacts were made from metal of meteoric origin, such as some of Tutankhamun's ritual swords, a rare and precious material found in its elemental state only in meteorites and in very few deposits. The Hittites built their empire due to their monopoly on ironworking technology, but when this spread, the relative abundance of iron led to a true 'democratization' of the metal, radically changing social and military structures.

The Iron Age marks the transition from prehistory to history. Today, iron and steel production is responsible for 30% of all industrial CO_2 emissions, as well as SO_2 depending on the ore used.

Iron is not only an industrial material but is essential for living organisms as a constituent of haemoglobin. Magnetite serves as a biological sensor of the Earth's magnetic field for carrier pigeons and other animals, allowing them to navigate during migration, as well as a dating tool for ocean floors, which has enabled geologists such as Wine and Matthews to provide the scientific basis for the theory of plate tectonics.

Lead: a dangerous protagonist in human history

The Romans used lead for pipes because it was soft and malleable, unlike iron, which was hard, reserved for weapons, and prone to rust. The toxic effects of lead were known to the Romans but were underestimated in favour of its practicality.

Lead causes saturnism, a disease widespread in ancient Rome that caused infertility, anaemia, abdominal pain, and nerve and kidney damage. At least nine emperors are believed to have suffered from it: Tiberius, Domitian, Caligula, Commodus, Claudius, Nero, Nerva, Trajan, and Hadrian. Roman practices increased exposure: wine was distributed through lead pipes; lead diacetate produced 'Saturn's sugar'; and the low pH of wine promoted high concentrations of lead.

Cerussite produced *ceruse*, a white pigment used by painters since ancient times. In the 1800s, it was used in “anti-aging” cosmetic creams: the skin remained young longer, hence Elizabeth I Tudor's “Youth Mask,” but the consequences were nausea, headaches, and paralysis.

Galenite was used in early radio receivers: a galena crystal allowed the construction of a diode, which was unreliable but inexpensive.

Gold: splendour and devastation

According to recent studies, gold and platinum arrived on Earth through a meteorite bombardment that occurred 4 billion years ago and lasted 200 million years. In its native state, even after separation from the rock, gold is always present in extremely low concentrations, usually associated with silver, copper, lead, zinc, and other metals. To produce one ounce of gold (approximately 31 grams), it is necessary to process an average of more than 100 tons of rock and produce 3 tons of waste.

To be marketable, gold must be at least 99.5% pure, but its processing is among the most polluting, both in terms of energy consumption and the production of highly toxic waste. The main methods of extraction are **cyanidation**, in which the minerals are treated with a solution of potassium cyanide, a highly toxic substance that is lethal to humans even at a concentration of 2%, and **amalgamation**, in which the minerals are pulverized and treated with mercury to form an amalgam from which the gold is then distilled.

Mercury causes serious health problems: around ten million people suffer from diseases caused by mercury released into the air and water, which spreads through the food chain, particularly in fish such as tuna, swordfish, and cod. Due to its rarity, workability, and inalterability, gold has historically been associated with a mythical time of prosperity and abundance, the Golden Age, which recurs in various ancient cultures. For the Romans, the 'Aurea aetas' corresponded to the 2nd century AD, the period of maximum prosperity of the Empire.

The impact of gold mining on the environment

In addition to its role in the financial market, gold is now present in all hi-tech devices (tablets, smartphones, PCs) in small quantities. Only recently have systematic procedures been initiated for its recovery from electronic waste (urban mining).

Rock salt: the mineral that built empires

Rock salt is not just a sodium chloride mineral, but an element that built empires and determined trade routes. The "salt roads" represent one of the earliest forms of globalization. Salt deposits played a fundamental role in the development of prehistoric European societies. Hallstatt in Austria represents the most emblematic case: an entire era (the Hallstatt culture, early Iron Age) is named after this site. Salt mines led to the development of complex underground mining techniques as early as the Bronze Age. In Walczak in Poland and Cardona in Spain, other important salt deposits also influenced local development. Control of these resources led to the emergence of powerful social elites, as evidenced by the wealth of grave goods. Furthermore, as salt was essential for food preservation, these centres became crucial hubs of long-distance trade networks, fostering the exchange of goods, technologies, and ideas between different European cultures of the Bronze and Iron Ages.

The cathedral of Real Monté, in the rock salt deposits in the province of Agrigento

The Republic of Venice based part of its economic power on its monopoly of salt in the Adriatic, while in France the salt tax was one of the causes of the French Revolution. The word 'salary' derives from the Latin 'salarium', the ration of salt given to Roman soldiers, testifying to the economic value of this mineral in ancient times.

Beyond these European examples, salt has shaped societies globally. The ancient Saharan salt trade routes connected Mediterranean civilizations with sub-Saharan Africa, with cities like Timbuktu prospering as salt trading centres. In China, salt monopolies financed imperial governments for millennia, while in pre-Columbian America, salt production sites in the Yucatan and along the Pacific coast became centres of economic and political power. The Japanese tradition of salt-making through seawater evaporation developed distinctive cultural practices still celebrated today. These global examples demonstrate how a single mineral resource, distributed differently by geological processes, has shaped human societies across all continents, creating parallel but culturally distinct relationships between geology and civilization.

In Italy, the rock salt deposits of the Sicilian salt pans, formed during the Messinian salinity crisis about 6 million years ago, have shaped the local economy and inspired literature: Giovanni Verga in "Malavoglia" describes the hardships of the salt workers, while Luigi Pirandello sets "Ciàula scopre la luna" in the Sulphur mines associated with the salt deposits, creating powerful metaphors for the human condition.

Similar literary traditions connect geology and human experience globally. Émile Zola's 'Germinal' depicted French coal miners' struggles, while Upton Sinclair's 'King Coal' exposed conditions in American mines. Chilean poet Pablo Neruda wrote extensively about copper mining in the Andes, and South African literature documents the human cost of gold and diamond extraction. These literary works, emerging from different geological contexts, share a common thread: they reveal how mineral extraction has shaped not just economies but human dignity, social structures, and collective memory across cultures.

Tin: the metal that forged the routes of ancient Europe

The tin belt, stretching from Cornwall to Bohemia, characterized by hydrothermal deposits linked to granite intrusions, was fundamental to the development of the Bronze Age in Europe. The geological distribution of tin deposits determined the first trade routes and stimulated the development of navigation for the transport of the mineral. The main mining centres (Cornwall, Erzgebirge, Brittany) became hubs of technological innovation, developing increasingly advanced mining and metallurgical techniques, leading to the formation of specialized mining communities and the development of long-distance trade networks.

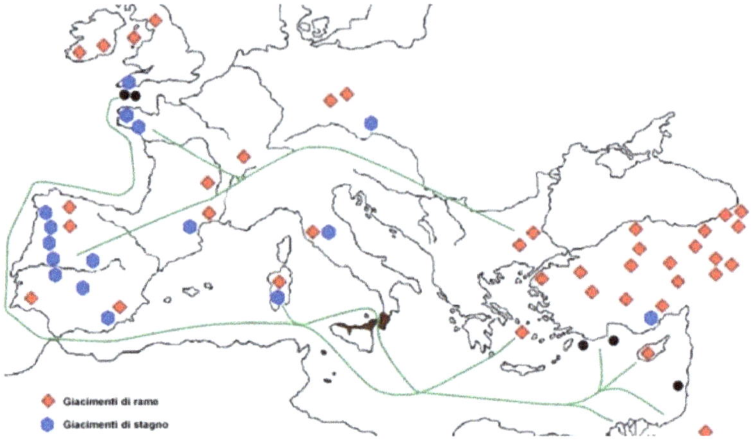

The Mediterranean area and the Copper and Tin deposits

Control of these resources influenced the balance of power and stimulated the development of technologies: the geological distribution of mineral resources shaped not only the economy but also the entire social structure of the first European civilizations.

Art and minerals: colours from the heart of the Earth

Minerals and art have been closely linked since prehistoric times. The earliest traces of the use of minerals for painting date back to the last ice age, between 15,000 and 35,000 years ago, in the caves of Chauvet, Lascaux, and Altamira, where the colours have remained almost unchanged for millennia.

Oxides and sulphides of metals or rare earths produced most of the pigments used in painting: yellow from goethite, red from hematite, ultramarine, blue, used for more precious details as it was obtained from expensive lapis lazuli, bright red from cinnabar, defined by Pliny as *Minio*, from which the term "miniate" used for medieval miniatures derives.

Red has always been a symbol of power throughout history: from the cardinal's robes to the prestigious "Pompeian red" that characterized the most refined homes of ancient Rome. This famous red was traditionally attributed to the use of cinnabar, a mercury sulphide mineral that produced an intense and brilliant hue. However, recent archaeological research has revealed a surprising discovery: many of the Pompeian houses were originally painted with yellow ochre pigments, and the characteristic red we admire today is the result of chemical transformation caused by the extremely high temperatures of the eruption of Vesuvius in 79 AD. The extreme heat would have altered the chemical composition of the original pigments, transforming the yellow into red through an unintentional 'baking' process that paradoxically created one of the most iconic colours in ancient art.

Colour metamorphosis in Renaissance art: from azurite to malachite

Of particular interest is the phenomenon of the transformation of azurite into malachite in Renaissance paintings, a chemical process that has altered numerous works of art over the centuries. Azurite was used by Renaissance painters as an inexpensive alternative to precious lapis lazuli for painting the mantles of the Madonna and the skies. This economical choice had long-term consequences that were not immediately apparent.

Unlike lapis lazuli, azurite was chemically unstable and reacted with moisture and carbon dioxide. Over the centuries, many paintings underwent this colour transformation, altering the artist's original intention and creating a discrepancy between how the works appear today and how they were conceived. This change manifested itself gradually, altering the perception of the works overtime.

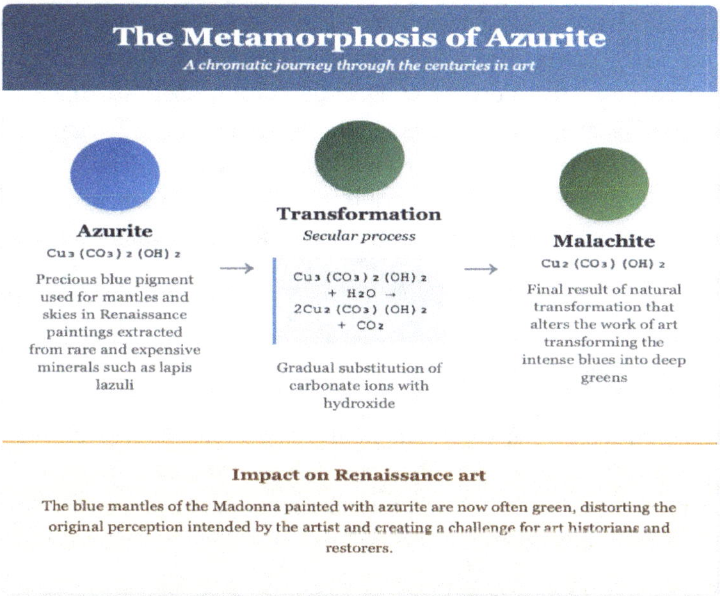

Some artists were more aware of this chemical issue. Michelangelo, for example, deliberately avoided using azurite in important frescoes such as the Sistine Chapel, preferring more stable pigments.

This transformation is an important example of the intersection between chemistry, mineralogy, and art history, and is crucial for restorers who need to interpret the original colours of works. Understanding these processes now allows us to reconstruct the original appearance of paintings and understand the artistic intentions of Renaissance masters.

Different shades also expressed social and economic power, as in Raphael's depictions of Pope Julius II. The transformation of dyes into pigments led to the great evolution of painting techniques, exploited by the Impressionists.

The invention of the metal tube in 1841 revolutionized painting: colours became transportable, allowing artists to paint outdoors and paving the way for Impressionism.

The relationship between geology and art extends across all artistic traditions. In Indigenous Australian art, ochres—iron oxide minerals—have been used for over 40,000 years, with specific ochre deposits holding sacred significance and trading networks extending across the continent. Japanese artists developed sophisticated techniques using mineral pigments including azurite (iwaenogu), creating the distinctive colours of traditional painting. In Islamic art, lapis lazuli imported from Afghanistan's Badakhshan mines created the brilliant blues of Persian miniatures and Ottoman Mosque decorations. Chinese porcelain, dependent on kaolin clay deposits, developed into an art form that influenced global aesthetics and sparked European quests to replicate its geological secret. These examples demonstrate how regional geology has influenced not just the availability but the very aesthetics of different artistic traditions.

Rare earths: when geology becomes geopolitics

Rare earths represent a perfect meeting point between geology, technology, and geopolitics. Elements such as neodymium, dysprosium, and cerium are essential for permanent magnets in wind turbines and electric motors, for catalysts, and for screens in electronic devices.

Despite their name, these elements are not rare in the Earth's crust, but they are rarely found in economically exploitable deposits. China dominates the world market thanks to the Bayan Obo deposit (the "Mountain of Rare Metals"), with profound implications for global economic and political balances.

Coltan, the mineral from which tantalum is extracted, is at the centre of conflicts in the Democratic Republic of Congo, highlighting the link between geology, resource exploitation, and human rights.

Energy transition minerals: contemporary challenges

The transition to a low-carbon economy has brought minerals that were considered marginal until a few decades ago to the forefront of global attention. These 'critical minerals' are a perfect case study for linking geology, technology, sustainability, and geopolitics in an integrated educational approach.

Lithium, a key element in rechargeable batteries, is mainly extracted from salt flats in Argentina, Bolivia, and Chile—the 'lithium triangle'—or from pegmatite rocks in Australia. Growing demand for electric vehicles has transformed this element from a chemical curiosity into a strategic global resource, with profound environmental and social implications. The process of extraction from salt flats, which requires the evaporation of huge amounts of water in arid regions, raises questions about the sustainability of technologies presented as "green."

Cobalt, another essential component of lithium-ion batteries, comes from the Democratic Republic of Congo for more than 60%, where extraction often takes place in conditions that are problematic from a human rights perspective, with documented cases of child labour. This mineral exemplifies the concept of the "ethical footprint" of technological products, prompting reflection on the responsibility of consumers and companies. Nickel, traditionally associated with stainless steel production, has become essential for next-generation batteries.

Indonesia, with its nickel-bearing laterites formed by the alteration of ultramafic rocks in a tropical environment, has emerged as a dominant producer, shifting global geopolitical balances and raising concerns about deforestation. Copper, an element known since ancient times, is experiencing a renaissance in the age of electrification: an electric car requires up to four times more copper than a traditional vehicle, while renewable energy infrastructure is a major consumer. The geological distribution of copper deposits, concentrated in Chile, Peru, the Democratic Republic of Congo, and Zambia, influences global investment and the industrial strategies of major economic powers.

The ethical dimension of mining

Coltan is just the tip of the iceberg of a global issue that deserves further investigation. Conflict minerals—which also include tin, tungsten, and gold mined in war zones—raise key questions about human rights, the responsibility of technology companies, and consumer choices. Copper mines in Chile consume enormous amounts of water in regions already plagued by drought, creating conflicts with local communities.

Studying minerals opens discussions on ethics, economics, politics, and human rights, transforming them from abstract objects of study into powerful tools for developing students' scientific knowledge, civic awareness, and critical thinking about the complex interconnections between geology, the global economy, and social justice, preparing them to face the sustainability challenges of the future.

7.2 Rocks as pages of Earth and human history

By combining different minerals into complex structures, rocks allow us to read sequences of geological events and understand the processes that have shaped our planet, intertwining naturally with the cultural, artistic, and technological evolution of humanity worldwide. The examples that follow draw extensively from Mediterranean and especially Italian contexts—a region whose geological diversity and long human history provide rich illustration of rock-human interactions. However, every region possesses similar stories written in its own geological vocabulary. Teachers are encouraged to identify parallel narratives from their local geology: the sandstones that built their regional monuments, the volcanic rocks that shaped local landscapes, the metamorphic stones that defined architectural traditions in their area.

Rocks, like minerals, can be presented to students not as simple objects to be classified, but as witnesses to fascinating stories that connect the geological past to our everyday experience.

Igneous rocks: from the heart of the Earth to architecture

Igneous rocks offer an opportunity to recount both the internal dynamics of the planet and their use in human civilizations. Granite, for example, an intrusive rock formed by the slow cooling of magma deep underground, is also the material that has defined the architecture of civilizations from Ancient Egypt to modern monuments. The sculptures of Mount Rushmore and the pyramids of Aswan testify to the strength and durability of this material.

Basalt, the floor of the oceans, has provided fundamental evidence for the theory of plate tectonics. The columnar basalt of the Giant's Causeway in Ireland is not only a perfect example of the contraction of a lava flow as it cools but has also given rise to legends such as that of the giant Finn McCool, who is said to have built a road across the sea to Scotland. Similar formations can be found in Iceland and near Mount Etna, in the Alcantara River gorges, where hexagonal or pentagonal prisms are arranged in structures that appear to have been created artificially.

Porphyry, with its characteristic porphyritic texture, enjoyed prestige in ancient Rome, where it was considered a symbol of imperial power. Its intense red colour made it primarily suitable for columns, basins, and decorations on monumental buildings. The famous baptismal font in Milan Cathedral, carved from a single block of porphyry, is a masterpiece that testifies to the skill of medieval craftsmen in working this hard volcanic rock.

Obsidian, a black volcanic glass formed by the rapid cooling of lava, has been used since Neolithic times to make cutting tools thanks to its excellent conchoidal fracture. Obsidian arrowheads, blades, and scrapers represented advanced technology for many prehistoric cultures. In Sardinia, Monte Arci was the subject of extensive Mediterranean trade as early as 7,000 years ago, testifying to the existence of maritime trade routes well before the Bronze Age.

A porous black basalt produced by eruptions of Mount Etna has been used for centuries in traditional Sicilian architecture, defining the appearance of Catania, which was rebuilt after the eruption of 1669 and the earthquake of 1693. Its fire-resistant properties and local abundance made it the ideal material in a volcanically active region. Today, the craftsmanship of lava stone for decorative flooring and art objects still represents the cultural identity of the area.

Liotru lava stone, symbol of Catania

Rhyolite, the volcanic equivalent of granite, forms spectacular structures in Yellowstone National Park in the United States, where extensive rhyolitic lava flows are associated with thick deposits of volcanic ash of the same composition. The Neolithic monument of Stonehenge is partly made up of rhyolite blocks with unique acoustic properties, suggesting that the monument may also have been used as a sound space for ancient rituals.

Similar relationships between volcanic rocks and human civilization exist worldwide. In Mexico, the dark tezontle volcanic stone defines Aztec and colonial architecture in Mexico City, while Easter Island's moai statues were carved from volcanic tuff. Ethiopian churches in Lalibela are carved from volcanic rock, and Japanese castles often incorporate andesite from nearby volcanoes. In each case, local volcanic geology shaped distinctive architectural and cultural traditions, demonstrating how the Earth's igneous processes have provided both challenges and opportunities for human societies across diverse contexts.

Sedimentary rocks: archives of life and the environment

Sedimentary rocks are veritable archives of the planet's environmental history, recording millennia of the evolution of the Earth and life in their layers. Limestone, far beyond its simple chemical definition, is a 'living rock' formed by the accumulation of shells and skeletons of marine organisms. This connection between the mineral and biological worlds is evident in iconic monuments such as the pyramids of Giza, built from limestone blocks that were once part of ancient tropical seas, or in the white cliffs of Dover, a British national symbol made up of billions of microscopic skeletons of microorganisms.

Travertine, the star of Roman architecture in the majestic Colosseum, tells a different but equally fascinating story: a sedimentary rock formed by the precipitation of calcium carbonate in thermal waters, whose porous microstructures, visible to the naked eye, preserve traces of the activity of microorganisms that influenced the precipitation of the mineral. The same geological process created the spectacular white terraces of Pamukkale in Turkey, where geology transforms into a landscape of extraordinary beauty.

The travertine terraces at Pamukkale in Turkey

In the city of Bath in England, you can find an entire Georgian architectural complex built with the famous "Bath Stone," a yellowish oolite limestone characterized by small spheres of calcium carbonate formed in warm, shallow seawater. This stone, which has the unique property of being easy to work with after extraction and then hardening when exposed to air, was the ideal material for the expansion of cities in the 18th century, creating one of the most harmonious urban landscapes in Europe.

Sandstone, formed by the disintegration and subsequent cementation of pre-existing rocks, has enabled the creation of extraordinary monuments such as the temples of Abu Simbel in Egypt, carved from Nubian sandstone in the 13th century BC. The geological history of this rock is intertwined with contemporary history when, in the 1960s, these monuments had to be dismantled and moved due to the construction of the Aswan Dam, in an operation that demonstrates the complex interaction between geology, cultural heritage, and technological development.

In Petra, Jordan, the Nabataeans elevated the use of sandstone to an art form, exploiting the properties of this rock to create an entire city carved out of stone. The natural layering of the sandstone, with bands of different colours ranging from deep red to golden yellow to ochre, was incorporated into the architectural designs, creating natural decorative effects of extraordinary beauty. This technique demonstrates how ancient civilizations had developed a deep intuitive knowledge of local geology.

Base of a pillar strip in Pietra serena. Pazzi Chapel, Florence, Piazza Santa Croce

During the Florentine Renaissance, Pietra Serena, a grey-blue sandstone characterized by a fine, uniform grain, became the favourite material of Brunelleschi and other master architects. The contrast between this dark stone and white plaster defined the Renaissance aesthetic of the city, creating typical architectural forms that still characterize the historic centre of Florence today and define its urban identity.

Volterra alabaster, a finely crystalline variety of gypsum, links geology to art and craftsmanship: used since Etruscan times for cinerary urns and, more recently, for lamps and decorative objects. It owes its characteristic translucency to the crystalline structure formed during the Messinian period, about 5.6 million years ago, when repeated evaporation of the Mediterranean led to the deposition of enormous quantities of evaporites. This property of gently diffusing light has made alabaster a favourite material for art objects that seem to glow from within, creating a bridge between geology and artistic expression across the millennia.

Sedimentary rocks have defined architecture globally wherever geology and human settlement intersected. In India, sandstone built Mughal monuments including the Taj Mahal's red sandstone base. Australian colonial architecture utilized local sandstones with distinctive golden and red hues. In the American Southwest, Ancestral Puebloans built cliff dwellings from local sandstone, creating structures that have endured centuries. Argentina's limestone deposits provided building materials for Buenos Aires' European-style architecture, while Jerusalem limestone—a particular oolite variety—has been used continuously for over 3,000 years, giving the city its characteristic appearance. Each region's sedimentary geology created distinctive architectural vocabularies, demonstrating how local stone shaped built environments and cultural identities worldwide.

Nero's alabaster bathtub

Coal deposits tell a different story: formed by the accumulation and transformation of ancient forests into marshy environments, they bear witness not only to ecosystems that disappeared millions of years ago, but also to the industrial revolution that profoundly transformed human society. This fossil fuel, the main source of energy in the 19th century, enabled the industrialization of entire nations, changing landscapes, social models, and geopolitical balances. The "coal belt" in Central Europe and the eastern United States defined territories, local cultures, and even political movements, demonstrating how a rock can influence the course of modern history.

Coal deposits shaped industrialization patterns globally. Britain's coalfields powered the Industrial Revolution, while America's Appalachian coal fuelled westward expansion. Japan's limited coal deposits influenced its imperial expansion toward coal-rich regions. South Africa's coal supported both industrialization and apartheid-era economy. Chinese coal deposits enabled rapid recent industrialization but at enormous environmental cost. Indian coal continues to power development while creating sustainability challenges. Each region's coal geology influenced not just industrial development but political structures, labour movements, and environmental legacies that persist today.

Metamorphic rocks: transformations under pressure

Metamorphic rocks offer an opportunity to explore the concept of transformation, both in a geological and cultural sense. These rocks, formed by the transformation of pre-existing materials under the action of heat and pressure, tell stories of change that are intertwined with the evolution of human civilizations.

Marble, the result of limestone metamorphism, has shaped Western aesthetics from the Parthenon to Michelangelo's works. Its geographical distribution, abundant in the Mediterranean, has influenced the development of sculptural art in this region. Carrara marble, quarried from the Apuan Alps since Roman times, is the material of choice for sculpture. Michelangelo personally travelled to the quarries to select the blocks for his statues. The different qualities of this marble, from statuary white to veined Bardiglio, derive from the varying degrees of metamorphism undergone by the original limestone and the presence of mineral impurities. Giuseppe Sanmartino's Veiled Christ in the Sansevero Chapel in Naples is a masterpiece in which the sculptor's skill is combined with the technical characteristics of the marble's homogeneity and workability.

Beyond the Mediterranean tradition, metamorphic rocks have enabled distinctive sculptural and architectural traditions worldwide. In China, jade—a metamorphic nephrite—has held cultural significance for over 5,000 years, with jade carving developing into a supreme art form. In Zimbabwe, verdite and other metamorphic stones were carved into sculptures by Shona artists, creating a modern artistic tradition rooted in local geology. Pre-Columbian cultures in Mesoamerica prized green serpentine for ceremonial objects. In each case, local metamorphic geology provided materials that cultures invested with aesthetic, spiritual, and social significance.

Slate, derived from the low-pressure metamorphism of clayey rocks, has had a surprising impact on mass education: slate school blackboards have democratized access to education. The name derives from the town of Ardesio (Bergamo, Italian province)) where there were important quarries, and its properties gave the name to the town of Cavi di Lavagna in Liguria. Its use as a roofing material in northern Europe has shaped distinctive cultural landscapes.

Gneiss, a medium- to coarse-grained zoned metamorphic rock, can derive from different pre-existing rocks, telling complex stories and overlapping events. The banded arrangement of light and dark minerals represents the final stage of a metamorphic sequence that passes through clay schist, phyllite, and mica schist.

The variously deformed gneisses determine the formation of folds and structures that tell the story of the intense tectonic activity that shaped the Earth's crust.

Ophiolites represent fragments of ancient oceanic crust and upper mantle that have been tectonically emplaced onto continental margins during orogenic processes. These dark green metamorphic rocks—including serpentinites, metamorphosed gabbro, and transformed basalts—provide direct evidence of plate tectonic processes and ancient ocean basin closure.

In the Italian peninsula, ophiolites from the ancient Ligurian-Piedmontese ocean are extensively exposed in the Alpine and Apennine chains, offering exceptional opportunities to study these processes. Their presence tells the story of an oceanic crust subducted during the orogenic events that formed these mountain ranges, creating educational contexts where students can directly observe and touch remnants of oceanic lithosphere now found at high elevations in continental settings.

Serpentine extracted in the Alpine valleys of Piedmont and the Aosta Valley, dark green with white veins and known as "Alpine green marble," was considered by the Romans to be a material of extraordinary value. It was used in St. Peter's Basilica in Rome and in numerous Renaissance churches, demonstrating how Alpine metamorphic rocks have influenced sacred art.

Quartzite, derived from the metamorphism of quartz-rich sandstones, has been used since prehistoric times to produce tools thanks to its exceptional hardness. In Roman and medieval times, it was used for millstones, exploiting its resistance to abrasion. In the Alps, outcrops of white quartzite create landmarks visible from a great distance. The quartzite of the White Desert in Egypt, formed by the metamorphism of Nubian sandstones, creates a landscape of extraordinary natural geometry.

Among the rocks that have undergone various stages of transformation, flint deserves special mention. Formed from nodules of crystalline silica, it was the necessary material for cutting tools from the Palaeolithic to the Neolithic periods thanks to its conchoidal fracture. In the plains of northern Europe, it was later used in medieval architecture, creating churches with decorative facades featuring geometric patterns of black flint and white stones, as in the British tradition of "flint flushwork."

Unusual fretwork and flushwork at the Church of the Holy Trinity in Barsham, Suffolk

The same hardness that made it ideal for prehistoric tools was exploited in fire strikers until the 19th century.

Rocks and sustainable architecture: tradition and innovation

Contemporary architecture is rediscovering traditional stone materials in a new sustainable light, offering stimulating ideas for linking the study of rocks to the themes of environmental sustainability and technological innovation. Stone, a building material used for thousands of years, has surprisingly modern sustainability characteristics: exceptional durability, low embodied energy compared to concrete and steel, thermal insulation properties, and the ability to naturally regulate indoor humidity.

Its life cycle, from extraction to reuse, is an example of a circular economy ahead of its time. Architects such as Peter Zumthor have reinterpreted the use of local stone: the Vals Thermal Baths in Switzerland, built in grey-green quartzite quarried a few kilometres away, demonstrate how a geological material can create contemporary spaces in perfect harmony with the surrounding landscape, minimizing the building's ecological footprint.

Computerized cutting technologies have revolutionized stone processing, making it possible to obtain thin, lightweight elements that challenge the traditional perception of stone materials as heavy and massive. The Armadillo Vault, presented at the 2016 Venice Architecture Biennale, exemplifies this revolution: a self-supporting structure composed of 399 limestone blocks not bound by mortar or adhesives, whose complex shape, calculated using complex algorithms, perfectly distributes compressive forces.

The reuse of historic stone materials represents another frontier in sustainable architecture. The Punta della Dogana in Venice, restored by Tadao Ando, integrates 16th-century bricks and Istrian stone with contemporary interventions, creating a dialogue between different eras and reducing the consumption of new resources. These examples demonstrate how the study of rocks can be linked to highly topical issues such as sustainable architecture, cultural identity, and technological innovation, transforming what might appear to be a traditional subject into a window on contemporary challenges.

7.3 Teaching strategies for a narrative approach to "rocks"

The study of minerals and rocks can be transformed from a dry memorization of names and properties into an exciting journey through the history of our planet and human civilization. A narrative approach that connects "rocks" to everyday experiences, history, and culture can spark students' curiosity and make learning more meaningful.

Start with everyday objects

Starting the exploration of geosciences with familiar objects creates an immediate connection with students' lives. The technological devices we use every day are veritable portable mineralogical museums: a smartphone contains about 30 different mineral elements, including lithium in the batteries, silicon in microchips, indium in touch screens, and rare earths in speaker magnets, while a laptop has components derived from minerals mined in over 40 countries, creating a global network of economic dependencies.

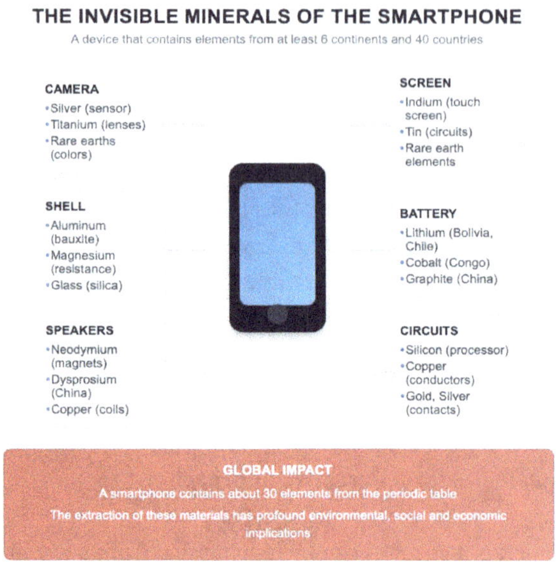

The tantalum used in capacitors comes mainly from the Democratic Republic of Congo, where its extraction is often linked to serious human rights violations. By following the path of these minerals from extraction to production, students can understand the global impact of their devices and the complex interconnections between technology, geology, and social justice.

School buildings themselves can become geological laboratories. Marble floors tell of ancient tropical seas and intense pressures, while granite staircases show the slow cooling of deep magmas. Plastered walls hide gypsum and limestone, witnesses to millennia of sedimentary processes.

Cosmetics offer another window on the mineral world: talc in body products, zinc oxide in sunscreen, mica in lip gloss, and white lead (basic lead carbonate) used as foundation by Roman nobles. By comparing these products across the centuries, students can discuss how scientific knowledge has influenced the evolution of cosmetic products, but also the role of women over time and beauty stereotypes.

Use "unexpected connections"

Unexpected connections between minerals and other disciplines create learning moments that give meaning to study. Art history intersects with mineralogy when considering pigments: changes in painters' palettes over the centuries reflect not only aesthetic evolution, but also the availability of minerals and trade routes.

Literature offers surprising connections to geology. Pirandello set his novella "Ciaùla scopre la luna" in the Sulphur mines of Sicily, where the protagonist, a caruso working in inhumane conditions, discovers moonlight for the first time upon exiting the mine, experiencing an epiphany that contrasts with his miserable existence. Similarly, Émile Zola in "Germinal" described the harsh conditions of coal miners. Exploring these texts alongside mineral samples creates a multisensory educational experience that combines science and humanity.

Music finds an echo in geology: quartz crystals oscillate at precise frequencies when subjected to pressure, a principle used in quartz watches. Marbles and sound rocks (lithophones) have also been used as musical instruments in different cultures.

Wars and historical events are often linked to mineral resources. The "Opium Wars" of the 19th century saw the United Kingdom trade Indian opium for Chinese porcelain (derived from kaolin, white clay). During World War II, the United States developed a secret program to acquire quartz crystals from Brazil, which were essential for military radio communications.

Integrating disciplines through minerals

Chemistry becomes tangible through minerals. Quartz crystals (SiO_2) illustrate covalent bonds, while the orderly cubic structure of table salt ($NaCl$) perfectly represents ionic bonding.

Physics finds applications in the behaviour of minerals. The birefringence of calcite, which makes images observed through a rhombohedron appear double, illustrates the optical properties of crystals. The magnetic properties of magnetite demonstrate magnetic fields, while piezoelectric quartz crystals generate electricity when compressed, a principle used in lighters.

Biology is linked to geology through biominerals: calcite and aragonite shells, hydroxyapatite bones, and fluorapatite teeth. Corals, by accumulating calcium carbonate in their skeletons, have created entire mountain ranges such as the Dolomites, once coral reefs in a tropical sea.

An extraordinary example of the interconnection between geology, oceanography, biology, and economics is represented by the phosphate deposits of Chile and Peru. These deposits were formed by a complex natural system that begins with the upwelling of the Humboldt Current, where nutrient-rich deep waters rise along the South American coast. This process feeds one of the most productive fishing grounds in the world, supporting huge colonies of seabirds. Their excrement, guano, accumulated over thousands of years in arid conditions, has formed extraordinarily rich phosphate deposits. In the 19th century, these deposits became so strategic for European agriculture that they triggered the "War of the Pacific" (1879-1884) between Chile, Peru, and Bolivia. A single mineral thus links oceanographic processes, marine ecosystems, climate, global agriculture, and geopolitical conflicts, offering an extraordinary example of the interconnectedness of Earth's systems.

Economic geography is best understood through the distribution of mineral resources. The African 'copper belt' in Zambia and the Democratic Republic of Congo determined colonial borders and continues to influence international relations. Phosphate deposits in Morocco and Florida support global agriculture, while Chinese rare earths dominate the production of green technologies.

The geology beneath our feet: valuing the local context

A particularly effective approach to teaching geosciences is to start with the surrounding area, transforming the local environment into an open-air laboratory where students can explore the connections between geology, landscape, history, and culture in a concrete way. Geological interpretation of the urban landscape is an accessible and enlightening activity: a walk through the streets of the city becomes a journey through geological time and space, observing the stone materials of buildings, monuments, and pavements. Investigating the origin of these rocks, the processes that formed them, and the reasons for their choice for specific applications transforms a historic building into a catalyst for scientific and cultural discoveries.

Place names often preserve valuable clues about local geological materials: street names, neighbourhoods, or localities that refer to quarries, mines, or distinctive geological features (Street of the Quarry, Slate Square, Kilns Quarter) are linguistic traces of ancient mining activities that shaped the local economy and society.

Traditional building materials tell the geological history of the area and its economic connections: Neapolitan yellow tuff extracted from the subsoil of the city of Naples, Pietra Leccese, which characterizes Salento Baroque, and Pietra di Vicenza, which dominates Palladian architecture, are examples of how human communities have developed construction techniques and architectural styles in close symbiosis with locally available geological materials.

Every region possesses equivalent examples waiting to be discovered. In North America, brownstone buildings tell stories of Triassic sandstone deposits; adobe structures in the Southwest reflect clay-rich soils; and granite monuments utilize glacially transported erratic. British cities showcase their underlying geology: Bath's oolite limestone, London's clay brick, Edinburgh's volcanic stone. Australian cities reveal local geology: Sydney sandstone, Melbourne bluestone basalt, Adelaide limestone. By investigating the stones beneath their feet—whether paving, monuments, or buildings—students anywhere can begin decoding their region's geological narrative and its influence on local culture and development.

Traditional productive activities linked to geological resources—from ceramics to bricks, from glassworks to metal workshops—have shaped the identity of entire communities. The rediscovery of these activities through educational programs that link geology to local craft traditions not only enriches scientific understanding but also promotes the enhancement of cultural heritage.

7.4 Towards a new perception of "rocks"

The narrative approach transforms seemingly inert elements into powerful tools for discovering the history of the Earth and humanity. Going beyond traditional teaching based on classifications, this perspective reveals the connections between geology and everyday life, stimulating curiosity about complex phenomena starting from familiar objects.

Through these stories, the ability to connect natural and human sciences develops naturally, promoting awareness of the temporal dimension of geological processes. The "rocks" thus cease to be dead objects and become witnesses to processes that connect the deep past of the Earth with the present and future challenges of humanity. Teaching minerals and rocks through their stories means helping students perceive this deep time and place the brief but intense human adventure on planet Earth within it.

Ropy basalt, in the Erta Ale caldera in Ethiopia

Further reading

McPhee, J. (1998). *Annals of the Former World*. Farrar, Straus and Giroux. https://us.macmillan.com/books/9780374518738/annalsoftheformerworld

Hjelle, A. (2022). Rocks and Minerals in Human History. Springer. https://link.springer.com/book/10.1007/978-3-030-83659-6

Kean, S. (2012). The Disappearing Spoon: And Other True Tales of Madness, Love, and the History of the World from the Periodic Table of the Elements. Back Bay Books. https://www.basicbooks.com/titles/sam-kean/the-disappearing-spoon/9780316051637/

Zalasiewicz, J. (2010). The Planet in a Pebble: A Journey into Earth's Deep History. Oxford University Press. https://global.oup.com/academic/product/the-planet-in-a-pebble-9780199645695

CHAPTER 8.
FOSSILS AS WINDOWS ON EVOLUTION: THE PALEONTOLOGICAL RECORD IN EDUCATION

Fossils represent an irreplaceable archive of the history of life on Earth, an archive that transforms evolution from a brilliant hypothesis into a theory confirmed by tangible evidence. Studying evolution without fossils would be like trying to understand an epic novel by reading only the last chapter: one might guess at recent dynamics, but the origins, fundamental conflicts, and countless transformations that shaped the story would remain invisible.

Although genetics, comparative embryology, and other disciplines offer valuable insights into the mechanisms of evolution, only fossils provide concrete evidence of past life forms and their succession over time. They are dual in nature: they are material objects, physical evidence of organisms that lived in remote eras, but they are also intellectual concepts that have guided and challenged scientific thought over the centuries.

8.1 Understanding deep time through fossils

One of the most difficult concepts to convey in teaching evolution is the vastness of geological time. The study of fossils offers a unique opportunity to make this temporal dimension, which eludes everyday human experience, tangible.

The arrangement of fossils in rock strata constitutes one of the first and most intuitive demonstrations of evolution. The vertical ordering of fossils, with simpler forms in older strata and progressively more complex forms in more recent ones, provides visual and concrete support for the notion of evolutionary change over time.

An effective educational approach is to present real stratigraphic sections, highlighting how guide fossils allow geographically distant rock layers to be dated and correlated. This not only illustrates the principle of faunal succession but also demonstrates how Palaeontology has practical applications in geology and resource exploration.

Stromatolites, laminated structures produced by colonies of cyanobacteria dating back 3.7 billion years, are a particularly effective example to illustrate the antiquity of life on Earth. The possibility of observing modern stromatolites in extreme environments such as Shark Bay in Australia also facilitates us to create a direct link between the past and the present of deep time less abstract.

Stromatolites in Shark Bay, Australia

Understanding deep time requires in fact a special effort and is not an intuitive concept for the human mind. As Stephen Jay Gould wrote: "Deep time is so difficult to grasp, so remote from our ordinary experience, that it can only be truly assimilated as a metaphor." Fossils help to construct these metaphors, making the immeasurable tangible.

8.2 Demonstrating evolutionary transformations through fossil sequences

Fossils offer the opportunity to directly observe morphological changes that have occurred over evolutionary time, providing concrete evidence of transformation processes.

Complete evolutionary sequences: success stories

Some evolutionary lineages are documented by exceptionally complete fossil sequences that convincingly illustrate the gradual transformation of species. These examples are powerful teaching tools:

The evolutionary sequence of horses, from the small Hyracotherium (the size of a dog) to modern horses, illustrates not only the increase in body size but also the evolution from multi-toed feet to single hooves in response to adaptation to a grassland environment. This sequence demonstrates that evolution does not necessarily follow a linear path but often proceeds through adaptive radiation with numerous side branches.

The evolution of whales from four-legged land ancestors is another remarkable example. Fossils such as Pakicetus (like a wolf with adaptations to the aquatic environment), Ambulocetus (the "walking whale") and Rodhocetus (with already reduced hind limbs) document the transition from land to a fully marine environment over the course of about 10 million years. This sequence demonstrates how anatomical structures can be re-proportioned for new functions.

The discovery of Tiktaalik, an intermediate organism between fish and tetrapod found in the Canadian Arctic, is a prime example of how evolutionary predictions can guide paleontological research. Researchers deliberately searched rocks of a specific age and environment and found exactly what evolutionary theory predicted: an organism with transitional features such as a fish body with a movable neck, robust ribs, and primitive "arms" with wrists and fingers.

Representation of Tiktaalik (credit: Ted Daeschler, Academy of Natural Sciences)

Transitional forms: bridging conceptual gaps

So-called "transitional forms" are effective in demonstrating evolutionary changes between large groups of organisms: Archaeopteryx, discovered just two years after the publication of Darwin's Origin of Species, is a classic example of an intermediate form between reptiles and birds. Its skeleton shows a combination of reptilian (teeth, long bony tail, claws on the wings) and avian (asymmetrical feathers suitable for flight) characteristics.

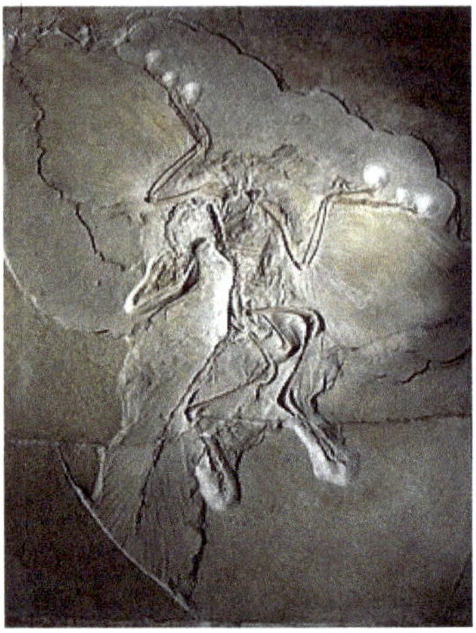

Archaeopteryx lithographica

A lesser known but significant example is Ichthyostega and Acanthostega, primitive tetrapod from the Devonian period that show transitional characteristics between fish and amphibians, including limbs with fingers but still adapted to a predominantly aquatic environment. These fossils demonstrate how evolutionary innovations often emerge in a functional context different from that which they would later assume: limbs with digits appeared before animals began to walk on land, but adaptation to the shoulder girdle was necessary for the limbs to be able to support a body and a skeleton in the absence of hydrostatic thrust.

It is necessary to clarify that 'transitional forms' are not organisms 'halfway' between two modern groups, nor do they represent links in a linear chain. Rather, they are real organisms, perfectly adapted to their environment, which shared characteristics that are now separated into distinct groups.

8.3 Exploring lost biodiversity and adaptive radiation

Fossils reveal an extraordinary diversity of life forms that have no equivalents in the present world, demonstrating how evolution has explored multiple adaptive solutions over time.

Explosions of diversity: evolutionary experiments

Particular periods in Earth's history have seen extraordinary adaptive radiations, in which numerous new life forms appeared in a relatively short time:

The fossils of the Burgess Shale fauna (Middle Cambrian, about 508 million years ago) reveal a surprising diversity of body plans, many of which have no modern equivalents. Organisms such as Hallucigenia, initially reconstructed upside down before being correctly identified as belonging to the group of onychophoran (velvet worms), or Opabinia with its five eyes and prehensile proboscis, demonstrate the extensive evolutionary experimentation that took place early in animal history. These examples introduce the concept of historical contingency in evolution: not all evolutionary innovations survive, regardless of their immediate effectiveness.

Opabinia - artwork by Hobo Tamura CC

Hallucigenia

Trilobites are another extraordinary example of adaptive radiation, with over 20,000 species documented in 270 million years of existence. Their diversity of forms, sizes, and ecological adaptations makes them excellent subjects for discussing the processes of specialization and diversification. In particular, the evolution of their compound eyes, with crystalline calcite lenses so perfect that scientists have been able to reconstruct how they saw the world, offers a fascinating example of the evolution of complex organs.

Mass extinctions: evolutionary resets

The major extinctions documented in the fossil record provide an opportunity to discuss how catastrophic events have repeatedly reshaped Earth's biodiversity.

Fossils allow us to study the five great mass extinctions, highlighting how different groups responded differently to environmental stresses. The end-

Permian extinction, which wiped out nearly 950% of marine species, demonstrates how global impact events can radically reshape the biosphere, creating opportunities for diversification among surviving groups.

An effective educational approach is to analyse extinction and recovery patterns, highlighting how dominant groups in one era can be replaced by previously marginal groups in the next. The extinction of non-avian dinosaurs at the end of the Cretaceous and the subsequent radiation of mammals clearly illustrates this principle.

When extinctions occur following datable and correlatability magmatic activity, they offer particularly interesting examples for reconstructing the complex interactions between geological and biological systems. The chain of events leading to these catastrophic events for ecosystems can be presented as a case study in environmental feedback.

These catastrophic events show how Earth's systems are interconnected. The massive volcanic eruptions in India during the dinosaur extinction released large amounts of gases into the atmosphere. This caused global warming and acid rain. These environmental changes caused a collapse of marine ecosystems, with ocean acidification and reduced oxygen availability. The fossil record documents this chain of events through sedimentary layers that show the progressive decline of planktonic forms and the disappearance of entire taxonomic groups, providing an extraordinary example of the interconnectedness between the geosphere, atmosphere, hydrosphere, and biosphere. Discussing these past events provides a historical context for grasp current environmental crises and their potential consequences for biodiversity.

→ intensification of magmatic, intrusive and effusive activity

→ increase in greenhouse gas concentration in the atmosphere
→ temperature increase
→ melting of permafrost present at high latitudes
→ dissociation of methane hydrates trapped in marine sediments
→ increase in CO2 concentration in the atmosphere
→ *positive feedback*
→ increase in ocean acidity (pH decrease) with negative impacts on calcareous shell organisms
→ *negative feedback*
→ expansion of tropical climate phases toward high latitudes and increase in precipitation
→ greater chemical-physical alteration of soils
→ greater erosion and surface runoff with greater nutrient input to lakes, rivers, and seas
→ emission of bio limiting and toxic metals
→ drastic increase in productivity
→ excess of organic matter
→ dissolved oxygen deficit
→ development of vast anoxic areas

8.4 Illustrating the mechanisms of evolution through fossils

Fossils not only document the results of evolution but also provide clues about the mechanisms that drive it, allowing us to illustrate the evolutionary forces at work.

Adaptation and natural selection

Numerous fossil examples clearly illustrate the process of adaptation to specific environmental conditions:

Woolly mammoths, thanks to the exceptional preservation of specimens frozen in Siberian permafrost, offer an ideal case study of adaptation to extreme cold. Their fur, which was up to a meter long, a 10 cm layer of fat, reduced ears, and other physiological adaptations (revealed by analysis of recovered DNA) demonstrate how natural selection can shape organisms in response to specific environmental pressures. Preserved stomach contents also allow us to reconstruct their diet and the environment in which they lived.

The miniaturization of island species, such as Homo floresiensis or Mediterranean dwarf elephants, illustrates how geographical isolation and limited resources can drive evolution towards smaller body sizes, demonstrating evolutionary plasticity even in groups normally characterized by large sizes.

Coevolution and ecological interactions

Fossils can reveal ancient ecological interactions, demonstrating how different species evolve in response to each other.

Fossil traces of predation, such as holes in mollusc shells or bite marks on vertebrate bones, document the ancient evolutionary "arms race" between predators and prey. The evolution of thicker shells in molluscs in response to predators with more effective mouthparts, or more sophisticated escape behaviours in vertebrates to evade faster predators, illustrate coevolution in action.

Plant fossils offer significant evidence of coevolution with pollinators. Fossil pollen grains, preserved in sediments dating back over 300 million years, document the evolution of reproduction in land plants and the transition from wind to insect pollination. Fossils from the Cretaceous period (145-66 million years ago) show the explosive radiation of flowering plants alongside that of pollinating insects, revealing one of the oldest and most significant interactions in the history of life.

Fossil amber preserves exceptional examples of these interactions 'photographed' at the moment they occurred: insects with pollen attached to their bodies, parasitoids with their victims, symbiotic fungi of insects, and flowers with specialized structures to attract specific pollinators. These Baltic amber fossils (approximately 40 million years old) allow us to discuss how ecological relationships represent powerful selective forces in evolution.

Not just adaptation: constraints and contingency

Fossils reveal that evolution does not always produce the 'best' solution but is limited by the history of organisms.

Starfish and sea urchins have maintained their five-rayed structure for over 500 million years, even though this shape creates functional difficulties. This shows how ancient characteristics can limit future changes.

An even clearer example comes from ichthyosaurs, extinct marine reptiles that closely resembled dolphins. Their tails looked identical to those of fish, but inside they had a completely different structure. In fish, the spine extends into the upper part of the tail, while in ichthyosaurs it extends into the lower part. This reveals their terrestrial origin: despite millions of years of marine evolution, they retained traces of their history as land-dwelling reptiles.

These examples show how evolutionary history imposes constraints on possible solutions, even when the environment would favour different forms.

Comparative anatomy: windows on evolution

Comparative anatomy is one of the most powerful tools for interpret evolution, and fossils are an essential component of it. By comparing anatomical structures between extinct and living organisms, students can directly recognise the principles of unity of type with modification.

A particularly effective activity is to provide students with homologous bones (such as forelimbs) from different vertebrates, both fossil and modern, and ask them to identify structural similarities and differences. Comparing the forelimb of a fossil reptile, a pterosaur wing, a bird wing, a bat wing, a whale fin, and a human arm reveals the shared structure (humerus, radius-ulna, carpal bones, metacarpal bones, phalanges) modified for different functions.

Fossil vestigial structures, such as the rudimentary hind limbs in primitive whale fossils or the vestigial pelvis in modern whales, offer compelling evidence of evolution. These structures, which have no current function but were present in ancestors, represent "evolutionary relics" that bear witness to an organism's evolutionary past.

The comparative anatomy of fossil hominin skulls allows us to trace changes in cranial capacity, the position of the foramen magnum, dental morphology, and other characteristics, documenting the gradual acquisition of human characteristics over time. By comparing the skulls of Australopithecus, Homo habilis, Homo erectus, and Homo sapiens, students can literally "touch" human evolution.

8.5 Human evolution: from the evolutionary bush to understanding ourselves

Human evolution offers a unique opportunity to explore complex evolutionary concepts through a narrative that directly touches on students' identities. Teachers should dispel a common myth right away: human evolution was not a direct evolutionary line from "primitive" to "modern."

Instead, the fossil record reveals a complex history with numerous branches, many of which became extinct without leaving direct descendants.

Today, we know that several species of hominins coexisted: Homo sapiens, Homo neanderthalensis, Homo floresiensis, Denisovans, and others. Ancient DNA analyses reveal that some of these species also interbred, creating a network of relationships much more complex than we imagined. These data allow us to discuss concepts such as geographic speciation and hybridization in a context directly relevant to our species.

Human evolution is an excellent example of mosaic evolution, in which different traits evolve at different rates and at different times. Upright posture, for example, significantly preceded the increase in brain size, as demonstrated by "Lucy" (Australopithecus afarensis), who walked on two legs but had a brain the size of a chimpanzee.

Fossils allow us to trace the gradual emergence of human characteristics: bipedalism, increased cranial capacity, reduction of the snout, changes in dentition, precision of grip, use of tools, and symbolic behaviour. These traits did not appear simultaneously, but through a complex sequence of adaptations to changing environmental and social conditions.

Human fossils, together with associated tools, allow us to explore the unique interaction between biological and cultural evolution. The increasing brain size in Homo fossils, accompanied by increasingly complex tools, illustrates how cultural and biological evolution influenced each other, accelerating both processes.

Fossil evidence of fire control, intentional burials, and primitive art offer windows into the evolution of cultural behaviour, allowing us to discuss the emergence of complex cognitive abilities and symbolic systems that characterize us as a species.

8.6 Fossils as the foundation of integrated evolutionary education

The incorporation of fossils into science teaching transforms these finds from static objects into powerful narrative tools that connect different scientific disciplines. Through fossils, evolution ceases to be an abstract theory and becomes a readable story, an extraordinary tale written in the rocks of the Earth.

Practical teaching strategies

Even a modest educational fossil collection is a valuable investment for any educational institution. Ideally, it should include common specimens that can be handled (ammonites, trilobites, shark teeth) and replicas of evolutionarily significant fossils. Practical activities can range from simulations of the fossilization process using common materials to fossil preparation workshops and palaeobiological reconstruction exercises where students deduce lifestyle and habitat from anatomical characteristics.

Excursions to local fossil sites offer irreplaceable learning experiences, contextualizing fossils in their sedimentary environments and developing responsibility towards the paleontological heritage. Before each outing, it is essential to check the regulations on fossil collection, turning this necessity into an educational opportunity on the protection of natural heritage.

Top-down approach and modelling

The teaching of Palaeontology lends itself perfectly to a top-down approach. Rather than starting with detailed taxonomic classifications, this approach begins with the major evolutionary processes visible in the fossil record: the emergence of multicellular life, the conquest of land, adaptive radiation, and mass extinctions. Next, the relationships between these events and the geological history of the Earth are explored, followed by specific case studies that illustrate fundamental evolutionary principles.

Physical models such as three-dimensional reconstructions of paleoenvironments and interactive simulations that visualize evolutionary changes are particularly effective in making the vastness of time tangible. Each fossil can be presented as a narrator of interconnected stories: the life of the organism, the process of fossilization, the ancient environment, and the history of its scientific discovery.

Integrated perspective

The fossil record illustrates the dynamic nature of Earth systems, the interconnection between environment and biological evolution, and the temporal dimension that puts our anthropocentric perception into perspective. By studying adaptive radiation and extinctions, students encounter the contingency of complex systems and the vulnerability of ecosystems in the face of rapid environmental change—lessons that transcend Palaeontology to touch our very awareness of being human in an ever-evolving universe.

TEACHING UNIT: Adaptive Radiation and Evolution

Title: "Evolutionary Explorations: The Case of the Burgess Shale Fauna"

Type: Interdisciplinary geo-bio teaching unit

Target: Secondary school (high school)

Duration: 5 hours (2 lessons + 1 workshop)

Approach: Top-down, from the macroscopic to the microscopic

Learning objectives:

Understand the concept of historical contingency in evolution
Analyse how fossils document evolutionary experimentation
Develop paleontological data interpretation skills
Integrate geological and biological knowledge

Lesson 1: Context and exploration (2 hours)

Presentation of the geological timeline with a focus on the Cambrian period (25 min)
Introductory video on the Cambrian "biological Big Bang" (15 min)
Group activity: analysis of images of Burgess Shale organisms (Hallucigenia, Opabinia, Anomalocaris) with hypotheses on habitat and way of life (50 min)
Comparison with current scientific interpretations (30 min)

Lesson 2: From observation to interpretation (2 hours)

Simulation of a paleontological excavation with models of Cambrian fossils (40 min)
Introduction to the concept of evolutionary contingency through the thinking of S.J. Gould (25 min)
Discussion: "What if we rewound the tape of life?" (25 min)
Analysis of the environmental conditions of the Cambrian and their influence on the explosion of biodiversity (30 min)

Final workshop (1 hour)

Creation of concept maps integrating geological and biological aspects (30 min)
Class discussion and interactive learning assessment quiz (30 min)

Suggested online resources:

Natural History Museum, London: https://www.nhm.ac.uk/discover/burgess-shale-fossils-and-their-importance.html
Royal Ontario Museum: https://burgess-shale.rom.on.ca/en/
3D models of Burgess Shale fossils on Sketchfab
Chengjiang Biota (China): UNESCO World Heritage Site with exceptional Cambrian preservation: https://whc.unesco.org/en/list/1388
Guizhou Province Paleontological Museum (China): Major collection of early animal fossils
Indian Statistical Institute (Kolkata): Research on Cambrian explosion from Indian subcontinent perspective
Various university collections worldwide maintain digital archives of Cambrian fossils accessible for educational purposes

Mono Lake—alkaline lake between Yosemite and Death Valley

Further reading

Benton, M.J. (2003). When Life Nearly Died: The Greatest Mass Extinction of All Time. Thames & Hudson. https://thamesandhudson.com/when-life-nearly-died-9780500291931

Dal Sasso, C. (2016). Italian Dinosaurs. Natural History Museum of Milan. https://www.museonaturalastoria.it/pubblicazioni/dinosauri-italiani/

Erwin, D.H. (2015). Extinction: How Life on Earth Nearly Ended 250 million Years Ago. Princeton University Press. https://press.princeton.edu/books/paperback/9780691165653/extinction

Alvarez, L.W., et al. (1980). "Extraterrestrial cause for the Cretaceous-Tertiary extinction." Science, 208(4448), 1095-1108. https://doi.org/10.1126/science.208.4448.1095

Gould, S.J. (1989). *Wonderful Life: The Burgess Shale and the Nature of History*. W.W. Norton & Company. https://wwnorton.com/books/Wonderful-Life/

Pievani, T. (2019). Imperfection: A Natural History. Raffaello Cortina Editore. https://www.raffaellocortina.it/scheda-libro/telmo-pievani/imperfezione-9788832851069-3139.html

Fortey, R. (2011). Survivors: The Animals and Plants that Time has Left Behind. Harper Collins. https://www.harpercollins.com/products/survivors-richard-fortey

CHAPTER 9.
COULD IT HAPPEN HERE TOO?
RAISING AWARENESS OF RISKS AND HAZARDS

The concept of "natural risk" needs to be thoroughly revised in education. The traditional approach to geological phenomena such as landslides, earthquakes, and volcanic eruptions presents them as exceptional, sudden, and unpredictable events, rather than as normal processes of the Earth system. This perspective generates a sense of powerlessness in the face of forces considered uncontrollable, hindering the development of a conscious relationship with the environment.

9.1 From "catastrophe" to systemic process

A systemic view offers a radically different perspective, based on an understanding of the interconnections that characterize the Earth system. Geological phenomena are recognised as natural components of Earth's evolution, integral parts of the processes that shape our planet. Landslides, for example, are fundamental processes in slope evolution that contribute to landscape shaping and material transport.

These phenomena are recurrent, with varying frequencies and intensities but often predictable within certain limits. Their recurrence can be analysed through historical and geological evidence, allowing the development of probabilistic models useful for land-use planning. Comprehending these dynamics is essential for developing effective adaptation strategies that adopt an integrated and preventive approach based on in-depth knowledge of the processes.

9.2 The triple nature of risk: hazard, vulnerability, and exposure

Risk is the result of the interaction of three factors, only one of which is directly related to natural processes.

Hazard is the probability that a natural phenomenon of a given intensity will occur in each period and area, characterized by intensity, frequency, spatial extent, and predictability.

Vulnerability represents the propensity of exposed elements to suffer damage, including structural dimensions (resistance of buildings), systemic dimensions (reaction capacity of social systems), and cultural dimensions (awareness of the population).

Exposure quantifies the elements present in the affected area: population density, economic value of goods and activities, presence of strategic infrastructure. It is directly linked to settlement and land use choices.

Risk can be conceptualized as R = f (H, V, E). Since vulnerability and exposure depend on human choices, the term "natural risk" is misleading. It would be more correct to speak of "risk from natural phenomena," emphasizing the ever-present anthropogenic component. Two cities in the same seismic zone may present drastically different risks depending on construction techniques and population density, demonstrating how risk is mainly determined by land-use planning choices.

9.3 Educational implications: Developing new educational approaches

As discussed in previous chapters, traditional teaching of geological phenomena often presents risks as exceptional events that are distant from students' everyday experience. This fragmented and decontextualized approach does not foster the development of a real awareness of the risks present in one's own territory. Instead, a systemic knowledge of geological processes requires innovative teaching methods that actively involve students and guide them to recognise the complex interactions between natural phenomena and human activities in the specific context of their territory.

Effective teaching approaches: Problem-Based Learning and Case Analysis

Problem-based learning (PBL) is particularly effective for studying geological risks because it encourages students to ask questions and seek solutions to real problems. This is not an innovative method, but its application to geosciences is especially relevant as it challenges students to deal with complex situations typical of natural phenomena.

In PBL, the learning process starts with concrete problems that students must analyse and solve. In the case of geological hazards, these problems may concern the stability of a slope, the safety of a building in a seismic area, or land-use planning in a hazard-prone area.

Several characteristics make PBL particularly suitable for the study of geological hazards. This approach encourages the formulation of open-ended questions rather than the search for pre-established answers, integrating theoretical knowledge and practical skills into a single learning process. It also promotes collaborative work and an interdisciplinary approach, developing critical thinking and the ability to make decisions under conditions of uncertainty.

A complementary approach is Case Analysis, which involves the detailed study of real cases of geological phenomena and their consequences. The study of cases such as historical landslides, seismic events, or volcanic eruptions allows students to analyse predisposing factors and triggers, environmental, social, and economic consequences, mitigation strategies adopted, and the effectiveness of preventive measures.

These teaching approaches promote a deep comprehension of concepts through their application in real contexts and develop cross-cutting skills such as spatial and systemic thinking, which are essential for interpreting the complex interactions of geological phenomena with the environment and human activities.

Teaching tools and models: from simplification to real complexity

Starting from Earth Learning Ideas, we wanted to develop a teaching model that introduced investigation, problematization, and more variables, creating contexts closer to reality. This approach requires students to recognise the different variables in geological phenomena while using simple and accessible tools. The goal is to stimulate an active learning process where students must independently identify the relationships between components in progressively more complex systems.

Experimental kits: evolution from simple tools to complex systems

The basic model proposed by Earth Learning Ideas is based on extremely simple elements: a tilting plane and a deck of playing cards. By observing the angle of inclination at which the cards begin to slide, students understand the principle of the angle of repose. Although effective for introducing a core concept, this approach is necessarily limited to a single variable (slope).

Starting from this basis, we wanted to develop a kit that would integrate the necessary variables to make the model more representative of real phenomena, allowing students to:

1. **Independently identify the relevant variables** in slope instability phenomena (not only slope, but also water content, substrate roughness, material type, etc.)
2. **Formulate hypotheses about the relationships between variables**, for example, predicting how an increase in water content will change the behaviour of different materials
3. **Design controlled experiments** by varying one parameter at a time to isolate the effects of each variable
4. **Collect and analyse quantitative data**, for example by measuring critical angles for different materials under different conditions
5. **Compare laboratory results with field observations**, linking simplified models to real phenomena observable in the field

This approach transforms a demonstration activity into a complete scientific investigation, where students must recognise and manage the complexity of natural systems. The materials remain simple and accessible, but the conceptual structure is significantly enriched, bringing it closer to the real dynamics of geological processes and the scientific investigation methodologies used by researchers.

9.4 The learning model and variables in the educational path

The innovative teaching approach for the study of geological hazards is based on a progressive learning path that starts with simple artifact and gradually leads to an understanding of the complexity of real phenomena. This method, developed from Earth Learning Ideas, transforms students from passive observers into active investigators.

The starting point is extremely simple tools: a tilting board and a deck of playing cards. By observing the angle at which the cards begin to slide, students understand the basic concept of the angle of repose. This basic model, while effective in introducing the principle, considers only one variable: slope.

Model for simulating the behaviour of different debris or grass cover with different slopes and substrates.

Model for simulating the behaviour of different rock blocks or layers, with low and high internal friction (smooth or rough layers), different slopes, and different substrates, from smooth to rough.

The educational evolution consists of progressively enriching the system with new variables that make it more representative of real phenomena. Students work with different materials (sand, gravel, clay), substrates with varying roughness, different water contents, and simulated vegetation cover. This increasing complexity enables for simple and accessible tools while exploring the richness of natural processes.

Through this approach, students develop structured research skills. They independently identify the variables relevant to instability phenomena, formulate hypotheses about the relationships between variables, and design-controlled experiments by varying one parameter at a time. They collect quantitative data by measuring critical angles for different materials under different conditions, then compare laboratory results with field observations, linking simplified models to real phenomena in the field.

Students learn to recognise and define key variables: substrate, material, presence of water, slope, and vegetation cover. This ability to identify components and processes within the system develops analytical skills. They must also specify the boundary conditions for each material and recognise interrelationships, using tables to organize information and draw generalizations.

The second phase of the training course involves moving from controlled experiments to the analysis of the real territory. Students use images, maps, GIS data, and aerial photographs to analyse concrete cases, verifying the conditions in which the phenomena occur and identifying the triggering factors. The aim is to develop the ability to recognise various phenomena in the familiar environment, reconstruct their history, and predict their evolution.

Key factors and risk communication

The success of this approach depends on effective classroom management, with the teacher acting as a guide and encouraging active student involvement. It is crucial to identify guiding questions that stimulate problem solving and lead students to compare the variables of the phenomenon with real cases in the local area.

An often overlooked but fundamental aspect is risk communication. Students learn not only to identify and analyse risks, but also to communicate them effectively through the creation of simplified maps, information campaigns, or simulations of briefings to authorities. The ability to translate technical concepts into clear messages is an essential skill for transforming scientific knowledge into effective preventive behaviour.

9.5 Towards a new culture of the territory

The transition from a catastrophic view to a systemic understanding of geological processes requires a profound cultural change. It is not just a matter of changing teaching methods, but of developing a new way of living in the territory.

History offers us valuable examples of communities that have been able to coexist with geological phenomena. Alpine populations have developed construction techniques to protect themselves from avalanches and landslides. The inhabitants of Vesuvius in ancient times recognised the signs of eruptions. Japanese architectural traditions have integrated seismic awareness for centuries. These experiences show that it is possible to develop a "culture of risk" based on knowledge and adaptation.

Education plays a central role in this change. The innovative teaching approaches described—from problem-based learning to practical activities with experimental kits—demonstrate how students can be actively involved. When students experience phenomena firsthand through experiments and direct observation, they develop a deeper and more lasting knowledge.

We need to overcome the traditional separation between "nature" and "human activities." Geological phenomena are not external events that occasionally affect our settlements, but processes with which our communities constantly coexist. Landslides, floods, and earthquakes are as much a part of the Earth's dynamics as our cities and our lives.

Only when geological awareness becomes part of our "sense of belonging" to the own region can we talk about true prevention. This is the most important contribution that geoscience education can make to building safer and more sustainable communities.

The DSGSD of Torgnon - A natural laboratory of geological risk

The following detailed case study, developed and tested in educational contexts in the Italian Alps, illustrates how the systemic approach and Problem-Based Learning methodology can be applied to slow-moving gravitational phenomena. While this specific example focuses on a site in the Aosta Valley, similar Deep-Seated Gravitational Slope Deformations (DSGSD) occur in mountain regions worldwide—from the Andes to the Himalayas, the Rocky Mountains to the Southern Alps of New Zealand—wherever tectonic uplift, glacial retreat, and gravitational processes interact. Teachers can adapt this framework to local examples of slope instability in their own regions, maintaining the same pedagogical structure while substituting regionally appropriate geological contexts and culturally relevant traditional knowledge.

The Deep-seated Gravitational Slope Deformation (DSGSD) of Torgnon is a concrete example of a phenomenon that is less immediately recognizable but has significant consequences for the morphology of the area and human activities. This case study offers the opportunity to apply the systemic approach and innovative teaching methodologies described above.

Located in the Torgnon Valley in the Aosta Valley, this DSGSD is a post-glacial deformation affecting an area of approximately 3 km² with an estimated depth of between 100 and 200 meters in the metamorphic substrate, mainly composed of gneiss and mica schist. The phenomenon has distinctive features such as multiple transverse steps on the slope surface, counter-slopes and closed depressions, trenches and fractures parallel to the line of maximum slope, splitting of the summit ridge and evidence of differential movement between blocks. The speed of movement varies from millimetres to centimetres per year, with accelerations during periods of heavy rainfall.

Related processes include the blocking of the main valley with the formation of a lake basin that is now filled, preferential flow channels that highlight the complexity of the movement, the formation of a paleo-fluvial plain following the blocking, and the development of water stagnation zones at the counter-slopes.

Torgnon-Aosta Valley

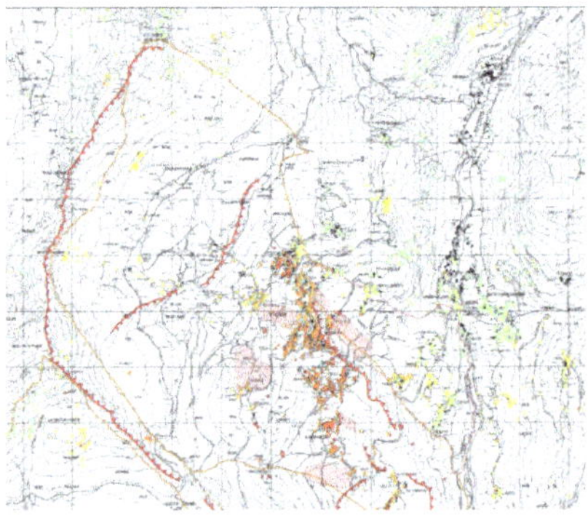

Mapping of instability

The Ru du Pan Perdu

TEACHING UNIT: The DSGSD of Torgnon

Target: Secondary school

Duration: 8 hours (classroom, laboratory, and field)

Approach: Systemic and Problem-Based Learning
Objectives:
· Understand DSGSD phenomena and their territorial impact
· Develop skills in reading and interpreting landscape signs
· Integrate scientific knowledge and traditional knowledge in risk assessment
· Develop proposals for mitigation and coexistence with slow-evolving phenomena

1. Overview of the phenomenon (2 hours) Analysis of distinctive characteristics: transverse steps, counter-slopes, closed depressions, fractures parallel to the slope, ridge splitting. Study of related processes: valley closure, lake basin formation, preferential flow channels, water stagnation areas.

2. Historical and cultural dimension (2 hours) Exploration of historical evidence: documents from 1600 on "walking land," the "Ru du Pan Perdu" (channel of wasted bread) and its centuries-old maintenance, significant place names such as "Rovine" and "Lou Plan dou Crépon" (plan of the crack). Collection of oral testimonies: stories handed down about morphological changes, collective memory of acceleration events, empirical knowledge of stable areas for building. Analysis of the evolution of risk perception from traditional communities to recent generations.

3. Practical educational activities (3 hours)

Document analysis: Comparison of historical maps and multi-temporal aerial photographs, identification of morphological changes, correlation with climatic events
Simulation: Creation of physical models of deformation mechanisms, simulation of the effects of different variables
Territorial survey: Mapping of morphological signals, interviews with the local population, analysis of traditional construction techniques

4. Summary and proposal (1 hour) Development of an integrated risk map combining scientific data and traditional knowledge. Development of mitigation strategies: proposals for territorial planning, structural and non-structural measures, communication strategies for the local community.

Skills developed: Landscape reading, systemic thinking, integration of scientific and traditional knowledge, multidisciplinary analysis skills, risk communication.

Interdisciplinary links: History (evolution of settlements), Geography (human-environment interactions), Social sciences (risk perception), Economics (cost-benefit assessment of strategies).

Final product: "Risk Atlas" of the DSGSD of Torgnon integrating scientific data, traditional knowledge, and management proposals, to be presented to the local community and relevant authorities.

This case study methodology—integrating scientific monitoring, historical documentation, traditional knowledge, and community engagement—provides a transferable framework applicable to gravitational hazards in diverse geological settings worldwide. The specific techniques for reading landscape signals, interpreting slow deformation, and developing community-based mitigation strategies remain valid regardless of location, while the content (geological formations, place names, local testimonies) would naturally differ in other contexts.

Yosemite National Park- Half Dome

Further reading

Peppoloni, S. (2021). *Science, Art and Philosophy: understanding and communicating natural risks*. INGV Edizioni. https://www.ingv.it/it/pubblicazioni-e-prodotti-editoriali/prodotti-editoriali/libri-e-guide/scienza-arte-e-filosofia-conoscere-e-comunicare-i-rischi-naturali

Guidoboni, E., & Valensise, G. (2013). *Italy and disasters: data and reflections on the impact of natural events 1861-2013*. Bononia University Press. https://buponline.com/prodotto/italia-dei-disastri-dati-riflessioni-sullimpatto-degli-eventi-naturali-1861-2013/

Luino, F., & Turconi, L. (2017). Floods and landslides in northern Italy in the period 2005-2016. CNR-IRPI. https://www.irpi.cnr.it/focus/eventi-di-piena-e-frana-in-italia-settentrionale-nel-periodo-2005-2016/

Crescimbene, M., La Longa, F., & Camassi, R. (2012). *Emergency psychology and stress management in organisations*. Quaderni di Geofisica, INGV. http://www.ingv.it/editoria/quaderni/2012/quaderno98.pdf:

Wisner, B., Blaikie, P., Cannon, T., & Davis, I. (2014). *At Risk: Natural Hazards, People's Vulnerability and Disasters* (2nd ed.). Routledge. https://www.routledge.com/At-Risk-Natural-Hazards-Peoples-Vulnerability-and-Disasters/Wisner-Blaikie-Cannon-Davis/p/book/9780415252164

Slovic, P. (2016). *The Perception of Risk*. Routledge. https://www.routledge.com/The-Perception-of-Risk/Slovic/p/book/9781853835285

UNDRR. (2022). *Global Assessment Report on Disaster Risk Reduction 2022*. United Nations Office for Disaster Risk Reduction. https://www.undrr.org/gar2022-our-world-risk

Peduzzi, P. (2019). *The Disaster Risk, Global Change, and Sustainability Nexus*. Sustainability, 11(4), 957. https://doi.org/10.3390/su11040957

CHAPTER 10.

DISASTERS ... AND OTHER STORIES. UNDERSTANDING PROCESSES, DISCOVERING UNEXPECTED CONNECTIONS

Geosciences are often perceived as "the science of disasters," and geologists as prophets of doom, Cassandras ready to predict dramatic events. In reality, they should be seen as interpreters of complex processes that deeply intertwine natural and human history.

The examples presented span all continents and geological epochs, demonstrating how Earth processes have influenced human societies across diverse cultural and geographical contexts. From African evolutionary processes to Mediterranean civilizations, Asian volcanic events to European climate shifts, and American indigenous histories to Oceanic colonization patterns, these case studies reveal universal principles: geological events operate on scales that transcend political boundaries, their effects propagate through complex environmental and social systems, and humanity's relationship with Earth processes constitutes a central dimension of our collective story.

In the previous chapters, we explored how to renew the educational approach to geosciences through contextualized modelling, narrative approaches, and a new awareness of risks. This chapter completes this vision by showing how geosciences represent a "bridge discipline" between natural phenomena and human affairs.

Geological processes are not isolated events, but forces that have given rise to new landscapes, social structures, and cultural movements, catalysing profound historical changes. From eruptions that inspired artistic masterpieces to earthquakes that redefined urban planning, from climate change that drove migrations to meteorite impacts that altered the course of evolution, geosciences offer a unique interpretative key to understanding the history of our planet and our species.

Through concrete examples and interdisciplinary connections, we will discover how seemingly distant geological events have influenced art, literature, politics, and society, transforming what might seem like a list of catastrophes into a compelling narrative of the interconnections between Earth and humanity.

Chicxulub: An Emblematic Case of Catastrophe with Unexpected Consequences

The Chicxulub meteorite is perhaps the most iconic example of how a geological catastrophe can have complex and far-reaching repercussions. This enormous celestial body, measuring 10-12 km in diameter, crashed into the Yucatan Peninsula 66 million years ago and was not simply an isolated destructive event, but the trigger for an evolutionary transformation of planetary proportions.

The impact sent a huge amount of dust and ash into the atmosphere, obscuring the sun for years and causing global cooling that drastically altered Earth's ecosystems. This "catastrophe" led to the extinction of the dinosaurs, organisms that had dominated the Earth for over 160 million years and which, contrary to the popular image of primitive creatures, had reached remarkable levels of evolution: many species were warm-blooded, some were ovoviviparous, many had developed feathers, and exhibited complex behaviours such as parental care.

But what was a dramatic end for the dinosaurs proved to be an unprecedented opportunity for the small mammals that had until then lived in the shadow of the large reptiles. Freed from predatory and competitive pressure, these animals were able to diversify, occupy new ecological niches, and evolve into increasingly complex forms—a process that, tens of millions of years later, would lead to the emergence of our species. This example perfectly illustrates the concept presented in Chapter 8: what appears to be a 'catastrophe' for some organisms represents an evolutionary opportunity for others, demonstrating the systemic nature of Earth processes.

The complexity of relationships, dynamics, and connections between geosciences and countless other scientific, historical, and cultural fields opens a multitude of educational examples. These connections, gathered from specialist and popular research, can become powerful tools for developing not only curiosity but also an investigative spirit in students, bringing them closer to geosciences through a journey of discovery built on logical steps. This approach reveals how geosciences have often been the trigger for events seemingly distant from this discipline, influencing the history of humanity and cultural developments, both past and recent.

When exploring the unexpected connections between geological events and human history, it is important to establish the level of certainty of these relationships from the outset. Some of these connections, such as the effects of the Tambora eruption in 1815 and the resulting 'year without a summer', are well documented through historical records, meteorological data and geological evidence. In these cases, the chain of cause and effect has been reconstructed with remarkable precision by scholars. For other connections, especially those relating to more remote periods of human history or prehistory, such as the eruption of the Toba super-volcan or the possible links between earthquakes and the decline of ancient civilizations, there is obviously less data and less certainty. However, effective learning requires to emphasize that even these hypotheses are not mere speculation, but theoretical constructs based on solid scientific foundations.

This dynamic of destruction and renewal characterizes not only meteorite impacts, but also other geological phenomena such as volcanic eruptions, earthquakes, and climate change, which have repeatedly reshaped not only natural ecosystems but also human societies. Volcanic eruptions have not only destroyed cities and altered landscapes, but have also influenced migrations, inspired artistic and literary masterpieces, and even contributed to technological evolution. Earthquakes have not only destroyed buildings but have also redesigned urban planning and stimulated architectural innovations. Climate change has led to the rise and fall of entire civilizations.

This journey through time and space requires an open mind, considering the Earth and its geological processes not as mere objects of scientific study, but as active protagonists in the great human narrative, capable of influencing our history in ways that are as profound as they are surprising. These are events, relationships, and connections known to many, especially geoscientists; it is therefore not our intention to amaze with " with special effects," but only to organize into a temporal, logical system, linked to the type of events, a set of information that would otherwise risk taking on a pedagogical, almost anecdotal aspect, certainly interesting but not very effective if the goal is to overturn the traditional approach to geosciences.

10.1 Volcanoes as architects of life on Earth

Volcanoes have played a fundamental role in the history of our planet, influencing both its physical structure and the development of life. Through their activity, they have contributed to the activation of plate tectonics and provided chemical elements and compounds essential for biological evolution.

Primordial volcanic eruptions released water vapor and carbon dioxide into the atmosphere, components that formed the basis of the Earth's atmosphere and created the natural greenhouse effect that made the planet habitable. Without this phenomenon, the Earth would have remained a frozen and inhospitable world.

Volcanic ash, settling over vast areas, contributed to soil fertility thanks to its richness in essential minerals such as phosphorus, potassium, and trace elements, creating fertile soils that supported the development of complex ecosystems. Particularly important is the role of underwater volcanoes, whose hydrothermal vents, with their extreme conditions of temperature, pressure, and chemical composition, may have been the natural laboratory where the first organic molecules and, potentially, the first forms of life on Earth were formed.

In addition to their impact on biological development, volcanic eruptions have also profoundly influenced human history, representing not only natural forces capable of destruction, but also catalysts for social, cultural, and artistic change. The eruptions of Tambora and Krakatoa offer perfect examples of how geosciences are linked to human history.

The cases presented illustrate how geological processes can be analysed using the systemic approach described in Chapter 9, following the complex chains of cause and effect triggered by a single event. From an educational point of view, they allow the narrative reconstruction method proposed in Chapter 7 to be applied, guiding students through the multidimensional consequences of a geological event in various fields: climate, agriculture, society, politics, art, and literature.

The Tambora Eruption: The year without a summer and its cultural consequences

The Tambora eruption is the best-known case of a "year without a summer" due to the complexity of the events triggered by its eruption in 1815. For this reason, I present it separately from a historical chronology, which will be followed in the examples that follow.

Between April 5 and 15, 1815, the Tambora volcano on the island of Sumbawa in Indonesia erupted, expelling large amounts of volcanic ash into the upper layers of the atmosphere. The eruption was 10 times more powerful than the subsequent eruption of Krakatoa; moreover, the Soufrière volcano on the island of Saint Vincent in the Caribbean in 1812 and the Mayon volcano in the Philippines in 1814 had already erupted abundant dust into the atmosphere, as evidenced by Sulphur compounds found in Greenland ice. These phenomena coincided with the

Dalton Minimum, a period of low sunspot numbers and consequent low solar activity from around 1790 to 1830, during which the sun is believed to have emitted less energy. At that time, the Little Ice Age was still underway, a period of general cooling of the planet that lasted from 1600 until at least 1850.

The year 1816 would therefore be defined as the 'year without a summer', with consequences that profoundly affected human history and culture. During this eruption, severe anomalies in the summer climate destroyed crops in Europe and North America. Populations were hit by a great famine; farmers faced severe hardship, and many livestock died. Storms, heavy rains, and subsequent flooding of major European rivers are attributed to the Tambora eruption.

These climatic conditions and widespread food shortages prompted many inhabitants of the Italian peninsula to emigrate to America in the hope of finding a better world: in October 1816, the authorities of the Duchy of Genoa noted a significant increase in requests for expatriation documents for the Americas.

The eruption of Tambora

However, when the migrants reached the American coast, they discovered that the weather conditions had caused poverty and famine there too, so they pushed westward. Many historians cite the year without a summer as the main reason for the 'conquest' of the American West.

The following year, a group of English writers engaged in the traditional Grand Tour southward of young British men stopped to spend the summer at Villa Diodati near Geneva, which had been rented by Lord Byron. Mary Shelley and John Polidori got together, along with other guests who, unable to spend their days on the lake due to bad weather, decided to challenge each other to write the scariest novel. Two of these stories became the most famous works of English Gothic fiction, notably Mary Shelley's Frankenstein and John Polidori's The Vampire, which was the first modern vampire story.

Children's literature was also indirectly influenced by the eruption of Tambora. The Brothers Grimm published the first complete edition in 1815, a collection they called “Children's and Household Tales,” many of which were set in cold climates.

The lack of fodder inspired Baron Karl Drais to seek new means of transportation without horses, inventing the Dandy Horse, the prototype of the modern bicycle.

The Dandy horse

Another historical event that took place in 1815 is worth remembering for its possible connection to the Tambora eruption and its climatic consequences: Napoleon's defeat at the Battle of Waterloo. His war strategy was to move troops quickly across the battlefield to defeat the enemy army, but in that rainy year, the night before the battle, numerous violent storms struck the countryside where Napoleon's troops were camped. The rain made the ground muddy and the French cavalry, generally Napoleon's winning weapon, was unable to move due to the mud, and the artillery was ineffective.

Painting was also affected: the high levels of ash in the atmosphere made the sunsets that year spectacular, with red and yellow colours, linked to the refraction effects produced by atmospheric dust, which can be found in the paintings of J.M.W. Turner. In 1832, the Babuyan Claro volcano in the Philippines continued to colour the skies of Turner's works with red and yellow.

William Turner's valiant Temeraire

In 1883, it was the eruption of Krakatoa, one of the most powerful volcanic explosions in modern history, audible thousands of kilometres away, that coloured the skies. The ash and gases released into the atmosphere created spectacular red-orange sunsets around the world for months. According to one intriguing theory, Edvard Munch was inspired by these dramatic skies to paint "The Scream" in 1893, capturing the anguish

The eruption of the Toba volcano: an evolutionary bottleneck for humanity

Between 75,000 and 70,000 years ago, beneath Lake Toba on the island of Sumatra, a super volcano erupted in what was probably the largest volcanic event of the last 25 million years, further cooling the planet's climate, which was already affected by the Würm glaciation. The catastrophic eruption produced 800 cubic kilometres of ash into the atmosphere, triggering a collapse in global temperatures. Perhaps one of the peaks of the Würm glaciation, called Tahoe, was a consequence of the Toba super-eruption. These enormous quantities of aerosols in the atmosphere reduced sunlight by up to 90%.

Molecular data attest to a decline in the population of Homo sapiens, an evolutionary bottleneck that saw a drastic reduction in population and then a restart from the few survivors of the cataclysm.

The human community in East Africa, one of the largest of the period, suffered the impact of the volcanic eruption, the consequences of which lasted several years and were so severe that they seem to have brought humans to the brink of extinction. According to a much-debated theory, the Toba catastrophe was primarily responsible for reducing the number of human species populating the planet to a few thousand specimens: Homo erectus began a slow decline that lasted about 40,000 years, mainly due to competition with Homo sapiens for food. H. erectus specimens were larger and stronger than H. sapiens, but less intelligent. Homo sapiens were able to adapt even to complex scenarios thanks to their ability to develop ideas and communicate them simply and effectively to their fellow humans. This 'bottleneck', reducing the number of individuals in the human population, may partly explain the low genetic variability in our species.

The Toba volcano, current image

The eruption of Flores and the disappearance of Ebu

Homo erectus disappeared from Asia about 30,000 years ago. But a group of survivors on the island of Flores in Indonesia may also have descended from him. The only fossil evidence of this is a woman named Ebu, who probably disappeared because of the eruption of the Flores volcano about 12,000 years ago.

The eruption of Thera and the end of the Minoan civilization

Around 1645 BC, although the dates are still uncertain, a catastrophic explosion of the Thera volcano, on the present-day island of Santorini in the Cyclades archipelago, led to the almost complete destruction of the island. Plato recounted the dramatic and sudden eruption, although it was probably preceded by several weeks of numerous warning tremors. The volcanic activity, which forced the population to flee Thera and seek refuge in Crete, began with a paroxysmal phase characterized by a rain of pink pumice, still visible on the island, followed by a subsequent explosion: a jet of compressed materials and superheated gases that reached the stratosphere.

Santorini today

The explosion triggered a tsunami in the Aegean Sea that devastated the coastal regions. Recently, a group of Greek geologists identified traces of those giant waves on the coast of Turkey, which spread across the Aegean Sea, reaching it in just two hours, while ash fell over a vast area, as far as Anatolia, hitting the ancient cities and causing climate change.

Santorini was a naval base of the Minoan Empire, and the date of Thera's destruction, with the resulting tsunami, seems to coincide with the gradual disappearance of the Minoan civilization, which prospered thanks to trade, and with the end of its supremacy in the Mediterranean, with the consequent rise of Mycenae.

The Lion Gate - Mycenae

The Mycenaean civilization was dominated by a warrior aristocracy that extended its power over Crete, the centre of the Minoan civilization, weakened by the eruption of Santorini. While some researchers argue that there is a significant gap between the natural disaster and the decline of the Minoan civilization, it is likely that the volcanic eruption and the resulting tsunami that struck the coastal cities, which were important ports, contributed to the collapse of the Minoan civilization.

The Santorini event also seems to coincide chronologically with some episodes narrated in the Bible, the 10 plagues of Egypt: volcanic ash, by reducing sunlight, caused the plague of darkness and acid rain; a consequent increase in rainfall and humidity may have favoured the development of locusts; the proliferation in the water of toxic algae, Oscillatoria rubescens, which when they die turn the water red, like blood, and oxygen-poor, may have forced the frogs to leave the water where they lived; the uncontrolled growth of flies and other insects may be linked to two other plagues: the death of livestock and ulcers on animals and humans.

Environmental conditions may have favoured the development of Claviceps purpurea, the so-called "ergot," in the rye used to make bread. Due to famine, bread was reserved for firstborn males, who were therefore the first to fall victim to it. Although scientifically plausible and consistent with the principles of volcanology, climatology, and biology, these hypotheses remain speculation based on temporal and geographical coincidences.

A depiction of the ten plagues of Egypt

1. Water in blood 2. The frogs 3. Mosquitoes 4. The flies 5. The death of livestock

6. Pustules 7. Hail 8. The grasshoppers 9. Darkness 10. The death of the firstborn

However, they perfectly illustrate how a single geological event can produce a cascade of effects that propagate through ecosystems, influencing human societies even at great distances from the site of the eruption, and even leaving traces in the narratives and myths of ancient cultures

The eruption of the Okmok volcano and the fall of the Roman Republic

In 43 BC, the eruption of the Okmok volcano in the Aleutian archipelago (Alaska) caused a sharp drop in global temperatures with exceptionally well-documented consequences in the Mediterranean area, where it profoundly influenced the social and political context of the civilizations of the time.

Okmok is an imposing shield volcano with a summit truncated by two overlapping volcanic calderas, formed during eruptions 12,000 and 2,050 years ago. The event of 43 BC, preceded two years earlier by a minor eruption, has been identified through the analysis of Arctic ice cores as one of the most violent eruptions of the last 2,500 years. The phenomenon occurred at a crucial moment: shortly after the assassination of Julius Caesar (March 15, 44 BC), at the beginning of a period that historical sources describe as characterized by unusually cold temperatures, famine, and widespread political instability.

Geochemical research on volcanic ash has definitively identified Okmok as responsible for this climatic anomaly. Data show that 43 and 42 BC were among the coldest years in the northern hemisphere in recent millennia, with average seasonal temperatures up to 7°C below average during summer and autumn. Precipitation increased dramatically: from 50-120% across southern Europe during the summer to 400% in the autumn.

The effects of the eruption are described in numerous classical sources. Virgil, in the Georgics, mentions the darkening of the sun after Caesar's assassination. Servius Honorius, a Roman grammarian, writes that "the light of the sun failed from the sixth hour until nightfall." Plutarch reports that the sun appeared "veiled and pale" and gave off so little heat that fruits shrivelled instead of ripening.

The social consequences were devastating. In April 43 BC, a severe crisis began in northern Italy, spreading to Greece in October 42 BC. Rome and other Italian regions suffered from famine, disease, and riots. Plutarch recounts that after their defeat at Mutina, Antony's army faced such hunger that the soldiers were forced to feed on wild fruit, roots, bark, and even animals they had never eaten before.

Egypt also suffered the effects of the eruption, which exacerbated the variability of the Nile floods, with repercussions on food supplies and the political stability of the Ptolemaic Kingdom.

Ides of March, the death of Julius Caesar

Caesar's assassination marked the beginning of a period of strife involving the Roman Republic and Ptolemaic Egypt, culminating in the rise of the Roman Empire. The climate crisis triggered by Okmok exacerbated social unrest, contributing significantly to this political transition for Western civilization, demonstrating how distant geological phenomena could have profound historical repercussions, both through direct climatic effects and through the symbolic interpretation of these events as omens in ancient societies.

The Darkness that shook the Middle Ages: submarine volcanoes and twenty years of winter

Starting in 536 ADS, the world experienced an extraordinary phenomenon: a period of darkness that lasted for almost twenty years. The sun appeared for no more than 4-5 hours a day, constantly veiled and with an intensity "very similar to that of the moon," according to contemporary accounts cited by Columbia University geologist Dallas Abbott.

This darkening caused dramatic global cooling and devastating famines that hit medieval populations hard. Chinese chronicles contain references to unexplained "eclipses" that have not been astronomically confirmed, while in the West, apocalyptic fears of the end of the world spread.

The origin of this phenomenon remained a mystery until recent research conducted by Abbott and John Barron. By analysing Greenland ice cores, the researchers discovered that powerful equatorial underwater volcanic eruptions were responsible for the anomaly. These eruptions vaporized enormous quantities of seawater, projecting calcium-rich sediments and microscopic tropical organisms into the atmosphere, the remains of which have been found perfectly preserved in the Arctic ice.

The peculiarity of this event lies in its origin: it was not caused by conventional terrestrial volcanoes. The surface eruptions documented during that period were not intense enough to justify such a prolonged phenomenon. The hypothesis of submarine eruptions, initially formulated due to the lack of traces in terrestrial sediments, was confirmed by chemical analyses of ice cores.

The enormous volume of material ejected created a persistent atmospheric film that reflected solar radiation for nearly two decades, producing a permanent semi-eclipse effect. The mark left by this period of darkness and abnormal cold, which lasted until 555 AC, is visible in the tree rings of the time, which show extremely reduced growth, bearing witness to one of the most severe climatic events of the last millennia, with profound consequences for the nascent medieval civilizations.

The eruption of Llopango and the decline of the Maya civilization

Llopango is a volcano located in present-day El Salvador, characterized by a large central caldera about 10 kilometres in diameter. Part of the Central American Volcanic Arc, it is one of its most active segments, with 20 other volcanoes in the region.

In 431 A.D., the volcano erupted with devastating force, destroying the territory within a radius of about 50 kilometres and scattering ash on a planetary scale. The eruption developed in stages: initially with pyroclastic flows alternating with an eruptive column 30 kilometres high, followed by the collapse of the caldera with enormous pyroclastic flows and a cloud of gas and ash that reached the stratosphere. Layers of ash several meters thick were deposited near the caldera, while neighbouring countries such as Guatemala, Honduras, Nicaragua, Costa Rica, and Mexico received deposits of several centimetres.

The precise dating of the event was confirmed by the discovery of ash fragments in a Greenland ice core, chemically identical to the products of the volcano. The Sulphur compounds, particularly sulfuric acid, altered the global climate, causing a cooling of about half a degree in the following years.

This catastrophe coincided with the early Classic Maya period (300-600 AD), a time of expansion for this culture throughout Central America. The eruption killed every human being within 50 kilometres of the volcano and depopulated the surrounding region for decades. Archaeological evidence indicates a sudden interruption in Mayan pottery production during that period, suggesting a significant impact on cultural development.

This case illustrates not only the resilience of human civilizations in the face of natural disasters, but also how geological events can influence the cultural and artistic development of societies: the interruption in ceramic production represented a discontinuity that may have altered the stylistic and technological evolution of Maya art. Research continues to reveal connections between this eruptive event and various aspects of Mesoamerican cultural development.

The Samalas eruption of 1257 AD and the medieval European famine

The Samalas volcano, located on the Indonesian island of Lombok, was a 4,200-meter-high colossus with a diameter of 8-9 km, part of the Mount Rinjani volcanic complex. The 1257 eruption ranks among the most violent in the last 7,000 years (VEI index = 7), eight times more powerful than Krakatoa (1883) and ten times more devastating than Tambora (1815).

The explosive event generated eruptive columns tens of kilometres high and pyroclastic flows that covered much of the island, even crossing the stretch of sea towards Sumbawa. The destruction was total for numerous settlements, including Pamatan, the capital of the kingdom of Lombok. The ash column reached a height of 43 km, causing incandescent avalanches of pumice and gas. The ash spread as far as Java, 340 km away, while the volcano deposited over 10 cubic kilometres of material. The sky remained dark for weeks, perhaps months.

The effects of the eruption were felt as far away as Europe. In 1258, crops failed to ripen due to the abnormal cold, and the ensuing famine decimated the European population, killing a third of London's inhabitants. Around 10,500 medieval skeletons (with estimates suggesting up to 18,000 bodies in total) have been found in Spitalfields Market, dated by radiocarbon to around 1250. Initially attributed to the plague or conflict, the absence of weapon wounds and scientific evidence revealed that this massacre was caused by the famine that followed the eruption.

The identification of Samalas as the culprit was only recently confirmed through radiocarbon dating, stratigraphic and geochemical analysis, dendrochronological studies, and the discovery of the "Babad Lombok," a 13th-century Javanese manuscript on palm leaves describing a "phenomenal eruption" lasting a week, with earthquakes and avalanches of burning material. This data has made it possible to date the eruption to between May and October 1257.

The Babab Lombok, text on palm leaves

A French monk from the Vosges Abbey, describing the summer of 1258, testified: "The sun's rays barely warmed the earth, clouds and rainy fog were so frequent that it seemed like autumn. The hay could not be harvested because of the incessant rain, the crops were destroyed and could only be gathered in September, but the seeds rotted in the barns."

The skeletons found show signs of malnutrition such as scurvy and rickets. Many victims suffered from hunger-induced diseases such as dysentery and from pathologies linked to the social breakdown caused by famine, such as typhoid fever.

The eruption of Laki on June 8, 1783, and the French Revolution

Lakagigar, also known as Laki, is a volcano consisting of a series of aligned vents located in southern Iceland. It is a fissure volcano that has produced two major eruptions in historical times, the first in 934 AD and the second between 1783 and 1784.

On June 8, 1783, after days of intense seismic activity caused by magmatic movements underground, a series of phreatomagmatic explosions, caused by contact between water and magma, led to the opening of a huge linear crater in the ground.

Along this fracture, 130 volcanic vents formed, producing lava fountains up to 1,000 meters high. In the early days, the eruption was violent and explosive but gradually became effusive, with fluid lava flows.

The linear volcano Laki, current image

During the entire eruptive episode, which ended in February 1784, 14 cubic kilometres of basalt erupted, invading a large area of the southeastern coast of Iceland, and volcanic ash reached a height of 15 kilometres, dispersing throughout the northern hemisphere.

The volcanic gases formed what was called the Laki fog over Europe, a toxic blue mist that remained suspended in the skies for months on end. The volcanic activity released 120 million tons of Sulphur dioxide and 8 million tons of gaseous fluorine into the atmosphere, which had dramatic effects throughout Iceland. The contaminated pastures killed more than 50% of the livestock on the island, causing a famine that killed more than 20,000 inhabitants.

Driven by winds, the toxic cloud reached Norway and Bohemia in mid-June, covering Berlin on the 18th, Paris on the 20th, and the coast of Great Britain on the 23rd, bringing stifling heat to London and its inhabitants. The Sulphur dioxide caused an increase in heat across the European continent, and the summer of 1783 was one of the hottest on record at that time.

Once the haze had partially dissipated, as with other volcanic eruptions, Laki changed the climate in a large part of the Earth, and the following winter was very harsh. Crops were lost, causing famine, hunger, and disease. The death toll in Great Britain was enormous: it is believed that the toxic aerosol caused at least 23,000 deaths, with another 8,000 added to this number who died from the cold.

After the heat recorded in the summer of 1783, the average temperature in the northern hemisphere dropped sharply by about three degrees, causing a decrease in the temperature difference between Eurasia and Africa and between the Indian Ocean and the Atlantic, limiting the ability of the monsoons to generate the rains that feed the rivers. In North Africa, the temperature rose by two degrees and the lack of rainfall meant that the Nile did not flood the land with its fertile waters, making it impossible to sow crops without the necessary irrigation. In France, periods of drought alternated with harsh winters and cold summers, contributing to the increasing poverty and hunger of the population. Severe flooding occurred in the French regions of Auvergne and Limousin, as well as in much of Germany, especially in the Lower Rhine region, because of heavy rainfall and the melting of snow accumulated on the mountain peaks. In 1785, the climatic conditions in France led to abundant harvests and a fall in product prices; in the following years, agricultural production was penalized by drought in the summer, excessive cold in the winter, and violent hailstorms that devastated crops. It was against this backdrop of famine that the Revolution broke out in 1789.

1783: The volcanic eruption that changed the climate -Iceland

These climatic changes were described and commented on in European newspapers, but no one made the connection with the eruption of the Laki craters except Benjamin Franklin, who, during a lecture given on December 2, 1784, demonstrated that it was fog from Iceland covering the skies of Europe that prevented the sun's rays from penetrating and causing the abnormal climatic conditions. As has since been proven, Franklin was right.

The connection between the Laki eruption and the French Revolution, while not the triggering cause, is yet another fascinating example of how geology can intertwine with history, helping to cause events that shape civilization.

10.2 Earthquakes as catalysts for social change

Earthquakes, like volcanic eruptions, are geological phenomena that have profoundly influenced the course of human history. Unlike volcanoes, however, earthquakes have always been perceived primarily as destructive events, making it more difficult to recognise their indirect "positive" effects: regardless of their intensity and perhaps even their location, considering the extent of human settlement on Earth, they are and will continue to be catastrophes. They may have different outcomes, depending on the attention that populations and their governments have paid to compliance with anti-seismic criteria for the construction of buildings; but some earthquakes, and their consequences, in particular tsunamis, will always have devastating effects.

In the case of earthquakes, it has been more difficult to discover unexpected connections, precisely because the dramatic nature of the effects prevails over any positive consequences. Positive examples include cases where, following a seismic event, regulations were formulated that required careful anti-seismic construction, but unfortunately, cases where these regulations have actually been complied with are rare: following the Calabrian earthquake of 1783, the then King of Naples, Ferdinand IV of Bourbon, appointed a Special Commission, headed by Francesco Pignatelli as Vicar General of Calabria, to carry out surveys, relief efforts, and reconstruction; unfortunately, all the recommendations of this commission were completely disregarded, as would happen in many other cases.

This pattern of inadequate implementation of seismic building codes following disasters is not unique to any region but represents a recurring global challenge. From reconstruction after the 1906 San Francisco earthquake to debates following the 2011 Christchurch earthquake in New Zealand, from post-earthquake reconstruction in Turkey to building practices in seismically active zones of Central America and Asia, the tension between scientific recommendations and practical implementation reflects universal dynamics of cost-benefit calculations, political will, cultural factors, and economic constraints that transcend specific national contexts.

The search for examples is still ongoing, but the connections already discovered are varied and interesting. As mentioned above, an attempt has been made to follow a chronological order, even though the dimensions and effects varied greatly in different cases.

The earthquake of Troy, 1300 BC

Around 1300 BC, Greece, along with the rest of the Aegean and eastern Mediterranean, was hit by a series of earthquakes that began around 1225 and continued until 1175. In addition, a significant climate change led to severe drought, resulting in major famines. Thus, in the entire eastern Mediterranean region, between the last years of the thirteenth century and the first decades of the twelfth, fertile lands were transformed into deserts.

Excavations at Hissarlik in western Turkey, the ancient Troy, as demonstrated by Schliemann's research, in which an international team of archaeologists has been working for years, have brought to light nine different cities, one on top of the other, and it is still unclear which of them was Priam's Troy. Research data has shown that in phase VI, around the 13th century BC, the city measured 80,000 square meters, with a population of about 5,000 inhabitants, who imported objects of considerable value from Mesopotamia, Egypt, Mycenaean Greece, and Cyprus. Troy was located between the Mycenaean world and the Hittite Empire, two of the great powers of the Mediterranean world during the Bronze Age. The conflicts between the two powers eventually led to a war that involved the city.

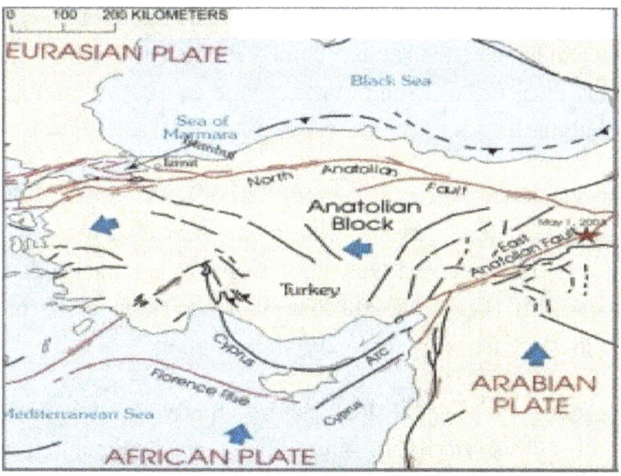

Troy and the geological-structural context

According to the most recent research, it was destroyed around 1330 BC. Troy was therefore not destroyed by invaders, but by an earthquake; perhaps it was too small to fully correspond to Homer's descriptions, although the earthquake may have facilitated its conquest by the Achaeans.

But then the city was rebuilt, and the seventh Troy was in turn destroyed around 1190-1180 BC by a fire following a siege, probably by the Sea Peoples and not by the Mycenaeans, who were already in a state of social crisis at that time.

The Noto Earthquake: Baroque Rebirth from the Rubble

Eastern Sicily is one of the most seismically active areas in Italy. The Iblei mountains, the northern portion of the African plate, collide with the Calabrian-Peloritano arc to the north, generating an accumulation of stress that is periodically released in strong earthquakes with long return periods.

The area has a long and dramatic history of earthquakes: as early as 1169, a violent earthquake killed at least 10,000 people and destroyed numerous towns. In 1542, another strong earthquake caused severe damage from Catania to Agrigento, affecting principally Noto, Lentini, and Syracuse, where the collapse of the cathedral caused numerous casualties.

But it was on January 11, 1693, that southeastern Sicily suffered its most devastating disaster, preceded by a foreshock on January 9 that had already damaged half of the buildings in Augusta and caused 200 deaths. The main shock reached an intensity of X-XI on the MCS scale, razing about seventy towns in the area between Catania and Ragusa, particularly in the Val di Noto. The earthquake was felt as far away as Calabria and Malta, with over 70% of buildings in Catania rendered uninhabitable.

The impact on the social structure was profound, with a total of around 60,000 victims: in Catania alone, at least 12,000 people perished, while Ragusa saw its population halved. The situation was made worse by a tsunami that affected the entire Ionian coast of Sicily for 200 km, with waves up to 10 meters high in Augusta and waters penetrating up to 200 meters inland.

Under the reign of Charles II of Habsburg, Sicily struggled to manage the emergency, with aid arriving very late. Cities were depopulated and survivors scattered across the countryside, afflicted by cold and malnutrition, which caused further deaths from epidemics.

However, the tragedy gave rise to an extraordinary opportunity: the Val di Noto underwent a radical reconstruction process that produced an architectural heritage of inestimable value. Giuseppe Lanza, Duke of Camastra, appointed general commissioner for reconstruction two months after the earthquake, led the area's rebirth. Noto was transformed into a "stone garden" thanks to the beauty of its new Baroque buildings. Catania was rebuilt according to a new urban plan with completely redesigned road layouts. Ragusa was partially rebuilt on the old site of Ibla and partly in its current location. In 2002, recognising the exceptional value of this architectural heritage born out of catastrophe, the Val di Noto was declared a UNESCO World Heritage Site.

The majolica of the staircase of Noto

The Lisbon Earthquake: A Philosophical and Urban Revolution

On November 1, 1755, Lisbon was devastated by a powerful earthquake with its epicentre of about 200 km off the Portuguese coast. The event generated a tsunami with waves up to 20 meters high that flooded part of the city and destroyed the Royal Palace and numerous buildings in the historic centre. A subsequent fire completed the destruction of the Portuguese capital. The effects of the earthquake also extended to the coasts of North Africa, causing 10,000 deaths in Morocco (concentrated in Fez, Tangier, and Marrakech), severe damage to Algiers, and devastation in Madeira, which was overwhelmed by the tsunami.

The disaster occurred in a city still heavily under the influence of the Holy Inquisition, which interpreted the event as "God's just response to the spread of heresy and lewdness." Some considered it divine punishment for the massacre of the indigenous people perpetrated by Portuguese soldiers in the Jesuit missions in South America. However, this dramatic event also marked the emergence of scientific and rational interpretations, generating a lively philosophical debate that gave rise to the "philosophy of disaster."

François-Marie Voltaire, one of the first theorists of European rationalism, openly challenged the religious interpretation in his "Poème sur le désastre de Lisbonne" (Poem on the Disaster of Lisbon): "[...] You will say, seeing these piles of victims: was this the price God made them pay for their sins? What sins? What crime have these infants committed, crushed and bloodied on their mothers' breasts? [...] Lisbon is destroyed and in Paris they dance."

This earthquake, which occurred in a century already marked by social tensions, contributed to a profound socio-cultural upheaval, transforming Lisbon from a city of medieval ecclesiastical obscurantism into the cradle of modern rationalism.

The Lisbon earthquake

Once the emergency was over, the Marquis of Pombal was tasked with planning the reconstruction of the city. Under his enlightened leadership, the old medieval alleys gave way to a modern, rational, and monumental city. The buildings were designed with innovative anti-seismic techniques, resting on solid wooden pillars. Their resistance was tested using scale models subjected to vibrations produced by trotting cavalry.

Pombal seized the opportunity to reduce the influence of the church and the aristocracy, promoting economic development and a new rational spirit. The collapse of religious power fostered a philosophical rethinking that would influence future scientific progress in Europe. Even today, the rebuilt Lisbon retains the imprint of this transformation, including the characteristic cobblestones made from the rubble of the earthquake.

10.3 Geological events and historical dynamics

The Earth is a planet in constant transformation, where extreme geological events mark the rhythms of natural and human history. Presenting extreme geological phenomena—volcanoes, earthquakes, meteor impacts—helps students understand the dual nature of catastrophic events: on the one hand, they bring destruction, but on the other, they are catalysts for profound change, reshaping entire ecosystems and altering the course of planetary evolution, opening new ecological niches and redefining global biogeochemical cycles.

This perspective extends to even larger-scale phenomena: Earth's magnetic field reversals, which cyclically expose the planet to cosmic radiation, altering climate and biodiversity; the Great African Rift, which is literally splitting a continent and redrawing geography and populations; super volcano eruptions that have influenced climate for thousands of years. Each of these processes operates on different timescales—from the seconds of an earthquake to the millions of years of tectonics—but all share the ability to rewrite the rules of evolutionary and cultural games.

Even more recent events such as the Tunguska (1908) and Chelyabinsk (2013) meteorites continue to remind us how thin the line between stability and change is. Studying these phenomena means not only discovering Earth's past but also interpreting the present and preparing for the future, connecting phenomena distant in time but united by their ability to radically transform the natural and human world in which we live.

The Tall El-Hammam Meteorite: Possible Origin of the Story of Sodom and Gomorrah

Meteorite impacts have left significant traces in human history. As early as 12,800 years ago, the prehistoric settlement of Abu Hureyra in Syria appears to have been struck by a small comet, as suggested by traces of molten glass and nano diamonds found at the site, making it potentially the oldest human settlement destroyed by a cosmic impact.

An even more fascinating case emerges from archaeological excavations in the ancient city of Tall el-Hammam (TeH), located at the southern end of the Jordan Valley between Lake Tiberias and the Dead Sea. This fortified urban centre, the largest Bronze Age city in the region, flourished for about 3,000 years from 4700 BC until its sudden destruction around 1650 BC. Significantly, TeH was not an isolated case: in the Jordan Valley, 15 other cities and more than 100 villages were abandoned at the same time, remaining uninhabited for 300-600 years.

The ruins of TeH show peculiar characteristics: the imposing 4-meter-thick walls appear to have been cut through, with the mud bricks mainly pulverized or reddened by fire, with no signs of natural erosion.

The ruins of Jericho

Archaeologists have discovered extraordinary evidence: quartz grains formed at high pressures, pottery with signs of melting (temperature above 1500°C), melted bricks and tiles (temperature over 1400°C), high concentrations of salt in the sediments, diamond-like pieces of carbon formed under high pressure, and iron- and silicon-rich spherules melted at temperatures above 1590°C. Exceptionally significant is the discovery of iridium in the sediments, an element rare on Earth but abundant in asteroids.

According to archaeologists, the only plausible explanation for this evidence was the explosion of a meteorite in the atmosphere a few kilometres from TeH. The event would have generated a very high-temperature heat wave that melted the exposed artifacts, followed by a shock wave that demolished and pulverized the structures, levelling the city and causing numerous casualties. The effects would have extended for tens of kilometres, explaining the simultaneous abandonment of numerous settlements in the region, including Jericho, which was repopulated only 300 years later. It is estimated that the population of the area collapsed from 40,000-60,000 inhabitants to a few hundred after the impact.

This catastrophic event, which occurred about 3,600 years ago, may be the origin of the famous biblical account of the destruction of Sodom and Gomorrah. The passage in Genesis describes the effects of a meteorite impact surprisingly well: "The Lord rained down burning Sulphur on Sodom and Gomorrah from the sky... He destroyed those cities and the entire valley and all the inhabitants... Abraham... saw smoke rising from the earth like the smoke of a furnace."

The magnetic field reversal

The Laschamps excursion, which occurred between 41,000 and 42,000 years ago, is an example of how reversals of the Earth's magnetic field may have profoundly influenced the history of our planet. During this event, the magnetic field—our natural shield against cosmic radiation—weakened and temporarily reversed.

Evidence of this phenomenon comes from kauri trees preserved in New Zealand's swamps, which show high levels of carbon-14 precisely at the time of the reversal. A weakened magnetic field would have exposed the Earth to higher levels of ionizing radiation, damaging the ozone layer and allowing more ultraviolet radiation to reach the surface. The highly ionized atmosphere would also have generated globally visible auroras and frequent electrical storms.

The consequences could have been dramatic: during this same period, Australian megafauna became extinct, and Neanderthals declined in Europe. At the same time, Homo sapiens rock art appeared with the first uses of red ochre, perhaps initially used as sunscreen—a practice that has survived to this day among some indigenous populations in Namibia.

The Oracle of Delphi

The site of Delphi, a World Heritage Site, was considered in ancient times to be the centre of the world due to the presence of the oracle, which influenced private and political decisions through the Pythia. Active from the 4th century BC until its closure in 392 A.D. at the behest of Theodosius, the oracle was already connected to geological phenomena according to ancient biographers: rock fractures, gas emissions, and springs.

While French excavations a century ago found no evidence of gas, recent studies have revealed that the temple stood exactly on a system of fractures connected to the Kerna Fault. Carbon dioxide, hydrogen sulphide, and methane rose through these fractures, especially during seismic phases. The springs also contained dissolved gases such as ethylene, benzene, and hydrocarbon mixtures.

Inhaling these vapours in the confined space of the sacred cell could easily induce trance states in the priestess. Plutarch described a "sweet scent" perceptible in the cell, consistent with ethylene, which, depending on its concentration, induces states ranging from euphoria to delirium and even loss of consciousness. Geology would thus explain the millennial mystery of the Oracle of Delphi.

The Rift Valley

Of great importance is the relationship between the formation of the Great Rift Valley, which began about ten million years ago, and the development of Australopithecus afarensis. This imposing geological barrier, six thousand kilometres long, blocked Atlantic disturbances, leading to a gradual drying up of the easternmost territories of the African continent. The process gradually fragmented the rainforest, replacing it with grasslands and savannahs—environments characterized by open spaces but also dangerous due to the presence of large predators. In this new environmental context, opportunistic bipedalism proved to be a decisive evolutionary advantage for anthropomorphic apes. Forced to move more and more frequently in open clearings, the upright posture offered multiple benefits: it reduced the body surface area exposed to the equatorial sun, allowed a wider view of the grassy expanses, which was essential for spotting predators, enabled running over long distances and, last but not least, freed the hands for other uses. These hands, equipped with opposable thumbs longer than those of other anthropomorphic apes, made a precise grip possible, paving the way for new manipulative abilities.

The documented fossil remains show a significant geographical distribution: Australopithecus afarensis (of which "Lucy" is the most famous specimen) is found mainly in the Rift Valley, while Australopithecus robustus is more common in deposits in the nearby highlands. This distribution suggests a differentiated adaptation to different environments: Afarensis, with its efficient body cooling, advanced manual skills, complex social structure, and mixed diet rich in animal protein, successfully adapted to the Valley environment, while Robustus probably followed a different evolutionary trajectory in the highlands.

Geological resources

The influence of geological resources on human technological development offers a relevant interdisciplinary perspective. The history of humanity is marked by access to and use of geological materials: from the Stone Age to the Copper, Bronze, and Iron Ages, human technological capabilities evolved according to the availability and ability to exploit specific mineral resources. The Industrial Revolution was made possible by the geographical coincidence of coal and iron ore deposits in regions such as England and Germany. In the contemporary era, the uneven distribution of rare earths and strategic minerals (such as lithium and cobalt) is shaping global geopolitical dynamics and influencing the pace of the energy transition. This interdisciplinary perspective allows us to understand how the "geological endowment" of different regions has had a decisive influence on trajectories of technological development, economic growth, and international relations.

Soils and agricultural civilizations

The impact of geological characteristics on the distribution of agricultural civilizations is another illuminating example. Soil properties, which are directly derived from the nature of the parent rocks and the geological processes that shaped them, have determined agricultural fertility and, consequently, the distribution of human settlements. The great river civilizations of antiquity—Mesopotamia, Egypt, the Indus Valley, China—developed in floodplains characterized by fertile soils, periodically renewed by flooding. Geological characteristics also influenced irrigation systems, agricultural practices, and even the social and political structures that evolved in response to specific local environmental conditions.

Climate change and human history

Climate change has always influenced the evolution of life on Earth, but only recently has science begun to understand how profoundly it has also shaped human history. Climate variations, regardless of their cause, have influenced the history of our planet, from the most significant ones such as Supergreenhouse and Snowballearth, which affected almost the entire Earth's surface, to more localized ones such as the Quaternary glaciations, up to the Little Ice Ages that have influenced more recent history.

In the past, the impact of these changes on historical events was often underestimated, as climate variations were not considered significant enough.

Today, however, cyclical or long-lasting climate phenomena have been decisive for many cultural and social developments in human history. Biological, dendrochronological, paleontological, and archaeological evidence reveals how climate fluctuations over centuries and millennia have repeatedly guided the course of human history, influencing migrations, the development or decline of civilizations, and even political and cultural events.

The migration of early humans from Africa

The evolutionary history of life on Earth has been constantly shaped by climate change. According to Darwin's theory, biological evolution represents a dynamic relationship between the genetic variability of organisms and changes in the natural environment in which they live. This is also true for the evolution of humanity, at least until geologically recent times, before we ourselves became the main cause of climate and environmental change.

One of the most significant episodes occurred between 900,000 and 800,000 years ago, when our African ancestors, like Homo erectus, were threatened with extinction due to prolonged drought caused by the drastic climate changes that marked the beginning of the Quaternary glaciations. This catastrophic event also had a generative impact, as it is linked to the emergence of a new human species: Homo heidelbergensis, the common ancestor of all subsequent human species, including Homo sapiens, Neanderthals, and the so-called "Denisovans."

African ancestors went through crises caused by cyclical climate change, with periods of higher humidity alternating with periods of greater aridity. The northernmost populations were subject to stronger environmental pressures, which probably selected those traits that led to brain development, social relationships, linguistic communication, and technological creativity.

Interestingly, some genetic studies suggest that a significant chromosomal event may have occurred around 800,000 years ago: the transition from 48 to 46 chromosomes, with the fusion of two pairs of ancestral chromosomes into a single pair (the current human chromosome 2). This important genetic modification could be linked to the selective pressures exerted by the climate changes of that period.

The 'Sahara Pump' and human migrations

The climate of the Pleistocene, with its characteristic alternation of glacial and interglacial phases, was a determining factor in the great migrations of the genus

Homo. In its migration from Africa, Homo sapiens crossed the Middle East and Anatolia and then reached Europe via two main routes: south of the Black Sea across the Bosphorus and Thrace, and east of the Black Sea across the Greater Caucasus Mountain range. The arrival of sapiens in Europe took place during a warm interstadial period of the Würm glaciation.

Subsequent population movements within Africa were influenced by alternating dry and wet phases in the Sahara and Sahel. During wet periods, these territories were transformed into green and fertile expanses crossed by waterways, attracting hominids. Conversely, during phases of desertification, they repelled these populations in all directions, creating a peculiar "pumping" effect and expelling them out of Africa.

This theory, known as the "Sahara Pump," explains that during wetter periods, North Africa and the Middle East were transformed into grasslands, allowing many different species to migrate from Africa to Asia. During drier periods, these areas returned to desert, interrupting this migratory corridor. The theory has been used to explain the migratory waves of various mammals, including horses, and of different human groups.

Sea level change: Doggerland and the Great Flood

Climate change has also caused major variations in sea level, radically transforming geography and creating or destroying migration corridors.

Doggerland was an area, now under the North Sea, that once connected the United Kingdom to continental Europe. It was a wide, shallow valley with low hills and marshes, rich in vegetation and wildlife, inhabited by bands of hunter-gatherers, as evidenced by finds of tools, prehistoric weapons, and animal bones. As the ice caps melted at the end of the last ice age, sea levels rose and gradually flooded the area. A tsunami linked to a sudden underwater landslide in the northern North Sea, around 8,200 years ago, may have accelerated this process. The rise in sea level appears to have permanently flooded Doggerland around 6,000 years ago.

The **Great Flood**, an event described not only in the Bible but in numerous documents of peoples who inhabited the Middle East as early as 8000 years ago, may be linked to a real event that can be located in the present-day Black Sea. During the ice ages, the area was covered by a large glacier that caused the ground to sink. When the glacier receded during an interglacial phase, a large freshwater lake formed. The continuous melting of the ice and the rising level of the

Mediterranean Sea eventually led to the breaking of the natural barrier between the lake and the Sea of Marmara, causing a massive flood that quickly transformed the lake into a brackish sea. This catastrophic event, which destroyed coastal settlements, may be the origin of the myth of the Flood. Remains of a human settlement have been found 91 meters deep in the Black Sea, dating back 7,500 years.

The Epic of Gilgamesh and the story of the universal flood

The Sahara: from green garden to desert

Between 7000 and 2000 BC, the present-day Sahara Desert had a completely different landscape: rich in rivers and lakes, green and marshy areas, where animals and humans lived. During the Neolithic period, the people who inhabited it produced pottery and gathered wild cereals. They had relations with pre-dynastic Egypt and scattered rock carvings throughout the Sahara. The rock art bears witness to the presence of numerous animal species that must have lived in the grasslands of the time.

Starting in 1200 BC, rainfall began to decrease, and gradually the Sahara turned into a desert, as described by Herodotus in the 5th century BC. The local populations tried to adapt, defending the few wet areas and building irrigation systems to safeguard the oases. Before the Roman conquest, there was still an organized society known as the Garamantes kingdom, which originated in the third millennium BC and declined until it disappeared in the 5th century AD.

10.4 Climate crises and transformations of ancient civilizations

The end of the Old Kingdom in Egypt (2300-2100 BC) coincided with a sharp reduction in the flow of the Nile, caused by a prolonged drought. Analysis of the relationship between oxygen isotopes in ice and the absence of lake sediments along the course of the Nile confirm this period of extreme aridity. The resulting social and religious revolution profoundly changed the role of institutions and religious beliefs. At the same time, the Achaeans from the north invaded Greece, probably in search of territories with more favourable environmental conditions.

The crisis of the great empires around 1300-1200 BC coincided with decades of low rainfall and widespread drought, as demonstrated by dendrochronology. The Egyptian and Hittite empires had to face attacks from the "Sea Peoples," probably driven by famine in their homelands. Social instability led to the collapse of large empires and small kingdoms such as the Mycenaean ones. From this crisis emerged the Greek culture that would characterize the Mediterranean before the rise of Rome.

Rome and the Empire: between 500 BC and 500 A.D., the climate in Europe was particularly mild, conditions that favoured the growth and expansion of the Roman Empire. The Roman climatic optimum (200 BC - 150 AD) allowed the colonization of Gaul and Britain, territories then characterized by a warmer and more fertile climate, where it was possible to grow wheat and vines. This favourable climate facilitated trade, commerce, and the spread of prosperity throughout the provinces of the empire.

The collapse of the Empire and the barbarians: the barbarian invasions were at least partly triggered by the climate deterioration that began around 250 A.D., when the global temperature dropped by about 1-1.5°C. This period, the Late Antique Little Ice Age (PEGTA, 400-750 AD), reduced agricultural production and drove populations such as the Visigoths, Ostrogoths, and Huns to migrate south in search of more productive lands. The waves of barbarian tribes that descended from the north towards the Mediterranean basin were fleeing territories that had become inhospitable due to frost and excessive rainfall.

The Little Ice Age and modern European history (PEG in Italy)

After the PEGTA (400-750 AD) and the Medieval Climate Optimum (1050-1270), characterized by a favourable climate that allowed agricultural prosperity, the development of communes and maritime republics, and even the colonization of Greenland, Europe entered the Little Ice Age (LIA, 1450-1850). In Iceland, during the Optimum, barley was even cultivated, and the mild climate allowed grapes to be grown as far north as Great Britain, a practice that disappeared with the onset of the LIA, leaving the French and Italians to dominate the wine market.

The LIG period was characterized by significant climate fluctuations: harsh winters followed by torrential rainfall in spring and summer, or moderate winters alternating with scorching summers and periods of severe drought. Exceptionally intense were the cycles of extreme cold between 1590 and 1610, when Europe experienced some of the lowest temperatures of the last millennium.

These adverse climatic conditions stimulated agricultural innovation, particularly in the Netherlands, with crop rotation that included fodder crops such as alfalfa, clover, and turnips. In England, the benefits of the new agriculture led to "enclosures," while in Holland, "polders" were created, land reclaimed from the sea using windmills.

These adaptations to the LIA gave rise to a different economic structure that contributed to the emergence of the Enlightenment and industrialization. The flow of gold from the New World accelerated the transition from a barter economy to a monetary economy, with the establishment of the Bank of England and the first modern banknote in 1695. In the 17th century, Europe "discovered" tulips and potatoes, and the landed aristocracy declined, replaced by a new class of bankers and merchants.

The difficult climate also influenced the cultural and religious spheres. Natural phenomena such as the aurora borealis, which was seen as far away as Switzerland in 1560, were interpreted as signs of the apocalypse. This climate of fear led to episodes of ethnic violence such as the pogroms against European Jewish communities, which between 1330 and 1500 suffered an unprecedented wave of anti-Semitism, accused of spreading diseases such as leprosy.

During the 17th century, the economic and social crisis worsened considerably. In 1648 alone, there were revolts in Russia and France, civil wars in Ukraine, Scotland, and England, and in the Ottoman Empire, the grand vizier was torn to pieces by an angry mob. Yet, paradoxically, this period of climate crisis coincided with extraordinary advances in the history of thought and science: the Renaissance, the Enlightenment, the American and French Revolutions, and the discoveries of figures such as Leonardo da Vinci, Galileo, Newton, and Linnaeus.

The causes of the LIA are still the subject of scientific debate. In addition to decreased solar radiation (the so-called "Maunder Minimum"), volcanic eruptions are thought to have played a decisive role, with unusual and intense tropical volcanic activity beginning between 1275 and 1300. Another theory considers the effect of the Atlantic Meridional Overturning Circulation: research by the University of Massachusetts has documented a change between the late 1300s and the early 1400s, when the waters south of Greenland and the Nordic seas became much warmer than usual, causing a rapid loss of Arctic ice that cooled the waters of the North Atlantic, diluted their salinity, and caused the cooling of the coasts of northern Europe.

One of the most striking effects of the LIA was the intensification of witch hunts. Starting in 1560, as the climate became increasingly unfavourable, people, especially single women, were accused of harming crops and animals, which were dying due to bad weather. The persecutions reached their peak in 1587-1588. Between 1580 and 1620, more than a thousand people accused of witchcraft were burned at the stake in the city of Bern, and in 1620, 900 women were convicted in Germany. The famous Salem witch trials of 1636 also took place in this context.

A fascinating side effect of the LIA was the extraordinary quality of Stradivari violins: the climatic conditions caused the slow growth of spruce trees, accentuating the density of the wood and its resonance. Wood from the Valcanale in Friuli was particularly prized for these characteristics.

While these examples focus primarily on European and Mediterranean responses to the Little Ice Age, similar climate-driven transformations occurred globally during this period. In China, the Ming-Qing transition (1640s) coincided with severe LIA conditions, contributing to agricultural failures and dynastic change. In Japan, the Maunder Minimum period saw famines and social upheaval. In North America, indigenous societies adapted agricultural practices and settlement patterns. In the Southern Hemisphere, Andean civilizations experienced their own climate challenges during corresponding periods. The LIA thus represents a truly global phenomenon with regionally specific manifestations, each revealing how different cultures adapted—or failed to adapt—to sustained climatic stress.

The LIA demonstrates how a relatively modest climate change (1-2°C on average) can have profound repercussions on every aspect of human society, from economics to politics, culture to religion, offering valuable lessons for understanding the possible consequences of current climate change.

10.5 Geosciences as a key to interpreting human history

This journey through the connections between geological phenomena and human history reveals the true educational potential of geosciences. The examples examined show that geological processes are not external forces that occasionally interfere with human civilization, but constituent elements of our evolutionary and cultural history.

The interdisciplinary approach developed here transforms the teaching of geosciences from a simple cataloguing of events to a narrative of the interconnections that have shaped our world. When students understand that the eruption of Tambora inspired "Frankenstein," that earthquakes have redefined architecture, or that climate change has driven great migrations, geosciences suddenly become relevant and engaging.

This perspective is crucial for addressing contemporary challenges. Current climate change, urbanization in earthquake-prone areas, and geological resource management are problems that require a systemic perception of the interactions between natural processes and human societies. Only by developing this awareness in students can we prepare them to face a future in which geoscience knowledge will be increasingly central.

Geosciences teach us that we are part of an ever-evolving planetary history. Understanding this history is not just an intellectual exercise, but a necessity for building more resilient and sustainable societies. From this perspective, every volcano, every earthquake, every climate change becomes a lesson about our relationship with the planet we inhabit.

The Connections between Geological Events and Civilization

VOLCANIC ERUPTIONS AND HISTORICAL TRANSFORMATIONS

PHENOMENON	WHEN	IMMEDIATE IMPACT	HISTORICAL CONSEQUENCES
Chicxulub Meteorite	66 million years ago	• Global darkening • Climate cooling • Mass extinction	• Evolutionary opportunity for mammals • Indirect origin of our species • K-T boundary in Palaeontology
Toba Supervulcano	75,000 years ago,	• -90% sunlight • Global volcanic winter	• Evolutionary bottleneck in H. sapiens • Decline of Homo erectus • Basis of low human genetic variability
Santorini (Thera)	~1627-1600 BC	• Island destruction • Mediterranean tsunami	• Decline of Minoan civilization • Possible origin of the ten plagues of Egypt • Rise of Mycenaean civilization
Okmok Volcano	43 BC	• Temperatures -7°C • Precipitation +400%	• Crisis in the Roman Republic • Contribution to transition toward Empire • Documented by Virgil and Plutarch
Underwater eruptions	536-555 AD	• 20 years of partial darkness • "Sun like the moon"	• Beginning of medieval "Dark Age" • European demographic crisis Agricultural and social transformations

PHENOMENON	WHEN	IMMEDIATE IMPACT	HISTORICAL CONSEQUENCES
Llopango	431 AD	• Central American devastation • 0.5°C cooling	• Interruption of Maya ceramics • Mesoamerican cultural discontinuity • Settlement abandonment
Samalas	1257 AD	• Ash to 43 km altitude • More powerful than Tambora	• 1/3 of London's population died of starvation • Volcanic winter in Europe • Scurvy and rickets epidemics
Laki	1783-1784 AD	• "Laki haze" over Europe • 120 million tons of SO_2	• Contribution to French Revolution • First scientific volcano-climate correlation • Famines in France
Tambora	1815 AD	• "Year without a summer" (1816) • Global climate anomalies	• Birth of "Frankenstein" and Gothic literature • Invention of the bicycle • Migration to American West
Krakatoa	1883 AD	• Power of 200 megatons • Spectacular sky colours	• Inspiration for Turner's sunsets • Influence on Munch's "The Scream" • Studies on optical effects of volcanic dust

EARTHQUAKES AND SOCIO-CULTURAL TRANSFORMATIONS

PHENOMENON	WHEN	IMMEDIATE IMPACT	HISTORICAL AND CULTURAL LEGACY
Troy Earthquake	~1300 BC	• Destruction of Troy VI • Part of regional seismic series	• Possible real basis of the Iliad myth • Facilitation of Achaean conquest • Transformation of Acgcan political landscape
Noto Earthquake	1693 AD	• 70 settlements destroyed • 60,000 victims • Ionian tsunami	• Sicilian Baroque renaissance • Radical urban transformation • Val di Noto UNESCO Heritage
Lisbon Earthquake	1755 AD	• 20-meter tsunami • Devastating fire	• Birth of "disaster philosophy" • Decline of the Inquisition • Enlightenment reconstruction • First modern anti-seismic techniques
Tall El-Hammam Meteorite	~1650 BC	• Temperature >1500°C • Destructive shock wave	• Possible origin of Sodom and Gomorrah • Abandonment of entire Jordan Valley • From 50,000 inhabitants to a fcw hundred

CLIMATE CHANGES AND HUMAN DEVELOPMENT

PHENOMENON	PERIOD	CHARACTERISTICS	HISTORICAL INFLUENCES
Laschamps Reversal	41-42,000 years ago,	• Weakening magnetic field • Increased cosmic radiation	• Neanderthal decline • Beginning of cave art • First use of red ochre • Development of human bipedalism
Rift Valley Formation	~10 million years ago	• Forest fragmentation • Savanna creation	• Evolutionary pressure for manual skills • Distribution of A. afarensis vs A. robustus • Migration waves from Africa
"Sahara Pump"	Pleistocene	• Alternating wet/arid phases • Temporary migration corridors	• Spread of technologies • Settlement distribution • Origin of flood myth
Marine Rise	8,200-6,000 years ago,	• Doggerland submersion • Black Sea flooding	• British genetic isolation • Migration toward interior • Roman Empire expansion
Roman Climate Optimum	200 BC-150 AD	• Milder temperatures • Greater agricultural productivity	• Mediterranean climate cultivation in north

PHENOMENON	PERIOD	CHARACTERISTICS	HISTORICAL INFLUENCES
Late Antique Little Ice Age	400-750 AD	• Temperatures -1.5°C • Agricultural production decline	• Commercial prosperity • Barbarian migrations southward • Pressure on late Empire • European demographic reconfiguration
Medieval Optimum	1050-1270 AD	• Favourable European climate • Longer summers	• Development of Communes • Greenland colonization • British viticulture • Crop rotation
Little Ice Age	1450-1850 AD	• Extreme winters • Rapid climate fluctuations	• Intensification of witch hunts • Unique quality of Stradivarius violins • Renaissance and Enlightenment flourishing

GEOLOGICAL PHENOMENA AND SOCIO-ECONOMIC CONSEQUENCES

PHENOMENON	EPOCH	CHARACTERISTICS	CULTURAL IMPACT
Oracle of Delphi	from 4th century BC	• Ethylene emissions • Gas from tectonic fault	• Influence on Greek political decisions • Scientific basis of religious practices • Altered states of consciousness of the Pythia
Resource Distribution	Various epochs	• Differentiated access to metals • Geographic coincidence of coal-iron	• Bronze Age, Iron Age • Industrial Revolution • Rare earth geopolitics
Soil Characteristics	Various epochs	• Differentiated fertility • Alluvial renewal	• Development of river civilizations • Specialized agricultural systems • Adaptive social structures

Notes on the methodology for reading connections

The connections between geological events and human history vary in the level of scientific certainty:

1. **Well-documented connections**: Events such as the eruption of Tambora and its effects on the climate, literature, and society in 1816 are widely documented through historical records, meteorological data, and geological evidence.
2. **Probable connections**: Relationships such as that between the eruption of Thera and the decline of the Minoan civilization, or between the earthquake

and the fall of Troy, are supported by significant archaeological and geological evidence but retain a degree of uncertainty.

3. **Hypothesized connections**: Some correlations, such as that between the Tall El-Hammam meteorite and the biblical account of Sodom and Gomorrah, represent scientifically plausible hypotheses that correlate geological events with historical or mythological narratives.

In all cases, these connections follow proven scientific logic and demonstrate how geological processes have repeatedly influenced and shaped the trajectory of humanity, sometimes in completely unexpected ways.

Old Faithful geyser - Yellowstone National Park in Wyoming

Further reading

Alvarez, L.W., et al. (1980). "Extraterrestrial cause for the Cretaceous-Tertiary extinction." Science, 208(4448), 1095-1108. https://doi.org/10.1126/science.208.4448.1095

Ballard, R.D., et al. (2000). "Deep-water archaeology of the Black Sea: The 2000 season at Sinop, Turkey." American Journal of Archaeology, 105, 607-623. https://doi.org/10.2307/507410

Bunch, T.E., et al. (2021). "A Tunguska sized airburst destroyed Tall el-Hammam a Middle Bronze Age city in the Jordan Valley near the Dead Sea." Scientific Reports, 11, 18632. https://doi.org/10.1038/s41598-021-97778-3

Carbognani, A. (2021). "Tall el-Hammam: the biblical city destroyed by a cosmic impact?". Media INAF. https://www.media.inaf.it/2021/09/22/tall-el-hammam-sodoma/

de Boer, J.Z., Hale, J.R., & Chanton, J. (2001). "New evidence for the geological origins of the ancient Delphic oracle (Greece)." Geology, 29(8), 707-710. https://doi.org/10.1130/0091-7613(2001)029<0707: NEFTGO>2.0.CO;2

Drake, B.L. (2012). "The influence of climatic change on the Late Bronze Age Collapse and the Greek Dark Ages." Journal of Archaeological Science, 39(6), 1862-1870. https://doi.org/10.1016/j.jas.2012.01.029

Fagan, B. (2004). The Long Summer: How Climate Changed Civilization. Basic Books. https://www.basicbooks.com/titles/brian-fagan/the-long-summer/9780465022823/

Foster, K.P., & Ritner, R.K. (1996). "Texts, storms, and the Thera eruption." Journal of Near Eastern Studies, 55(1), 1-14. https://doi.org/10.1086/373781

Maslin, M.A., et al. (2014). "East African climate pulses and early human evolution." Quaternary Science Reviews, 101, 1-17. https://doi.org/10.1016/j.quascirev.2014.06.012

Piccardi, L. (2000). "Active faulting at Delphi, Greece: Seismotectonic remarks and a hypothesis for the geologic environment of a myth." Geology, 28(7), 651-654. https://doi.org/10.1130/0091-7613(2000)28<651: AFADGS>2.0.CO;2

Pievani, T., & Cavalli-Sfoza, L. (2012). "Homo Sapiens: the great history of human diversity." MicroMega, 1/2012. http://telmopievani.com/2012/01/01/micromega-almanacco-della-scienza-12012-homo-sapiens-la-grande-storia-della-diversita-umana/

Ryan, W.B.F., & Pitman, W.C. (1998). Noah's Flood: The New Scientific Discoveries About the Event That Changed History. Simon & Schuster. https://www.simonandschuster.com/books/Noahs-Flood/William-Ryan/9780684859200

Sivertsen, B. (2009). The Parting of the Sea: How Volcanoes, Earthquakes, and Plagues Shaped the Story of Exodus. Princeton University Press. https://press.princeton.edu/books/paperback/9780691150215/the-parting-of-the-sea

Spiller, H.A., Hale, J.R., & De Boer, J.Z. (2002). "The Delphic oracle: A multidisciplinary defence of the gaseous vent theory." Journal of Clinical Toxicology, 40(2), 189-196. https://doi.org/10.1081/CLT-120004410

Wood, G.D. (2014). Tambora: The Eruption That Changed the World. Princeton University Press. https://press.princeton.edu/books/paperback/9780691168623/tambora

CHAPTER 11.
TRADITIONAL AND DIGITAL CARTOGRAPHY: DEVELOPING KEY SKILLS THROUGH READING THE TERRITORY

Cartography is one of the oldest and most sophisticated languages created by humankind to understand and represent the world. From the first rock carvings to today's geolocation applications, the ability to translate three-dimensional space into two-dimensional representations has accompanied the evolution of civilization. This translation has never been a neutral process: every map is the result of choices, simplifications, and priorities that reflect not only geographical knowledge but also cultural, political, and economic visions.

In the digital age, we are witnessing a radical transformation of cartographic tools. Paper maps have largely been replaced by digital representations, accessible from any device and continuously updated. Google Maps, Google Earth, GPS navigators, and countless geolocation applications have democratized access to geographic information, making traditional map reading and interpretation skills seemingly obsolete.

Analysing the implications of this transition from an educational and training perspective does not mean nostalgically contrasting the past with the present but rather analysing which specific skills are developed by different cartographic approaches, which are being lost in the digital transition, and which new skills are emerging in interaction with contemporary tools.

11.1 The languages of cartography: from paper to pixels

The legacy of traditional cartography

The paper map is an extraordinarily effective and concise communication system. A topographic map sheet from national cartographic institutes (such as the USGS in the United States, Ordnance Survey in the UK, IGN in France and Spain, or IGM in Italy) at a scale of 1:25,000 contains thousands of pieces of information organized according to precise codes: contour lines representing elevation, symbols indicating vegetation, roads, settlements, and water. This wealth of information is accessible only to those who possess the keys to interpreting the language of cartography.

Reading a topographic map means first understanding the scale, a fundamental concept that allows distances on the map to be related to real distances. It also means interpreting the meaning of symbols, recognising the shapes of the terrain through contour lines, and identifying the characteristics of the landscape through the elements represented. It is a process that requires specific knowledge, but also the ability to abstract and visualize space.

Paper's physical nature enables manipulation—folding, annotating, overlapping—which facilitates interaction with geographical information. A map is a document that unfolds in physical space with significant dimensions, allowing for an experience of "immersion" in the territory represented.

The Tabula Peutingeriana, named after Konrad Peutinger, who came into possession of it in the sixteenth century. III century A.D. Österreichische Nationalbibliothek in Vienna.

Planisphere by Juan de la Cosa, cartographer, explorer and pilot, the first nautical map to represent the coasts of the American continent – Naval Museum of Madrid [Public domain]

Map of the ocean floor -XX century made by Mary Tharp

The relevance of digital cartography

Digital cartography has radically transformed our relationship with geographical information. Maps have become interactive, zoomable, and customizable. We can instantly switch from satellite view to schematic representation, from road map to three-dimensional view. We can overlay layers of information, query specific points, and calculate optimal routes.

Google Earth has made a view of the Earth that was once reserved for astronauts accessible to everyone. Navigation apps guide us step by step to any destination, recalculating routes in real time. Geographic Information Systems (GIS) allow us to analyse, process, and visualize enormous amounts of spatial data, opening previously unthinkable analytical possibilities.

This abundance of tools has given nearly everyone access to geographic information, but it has also changed the way we interact with it. The mobile device screen with its small size frames limited portions of the territory, which does not allow for an overview, contextualization, or appreciation of the complexity of the environment.

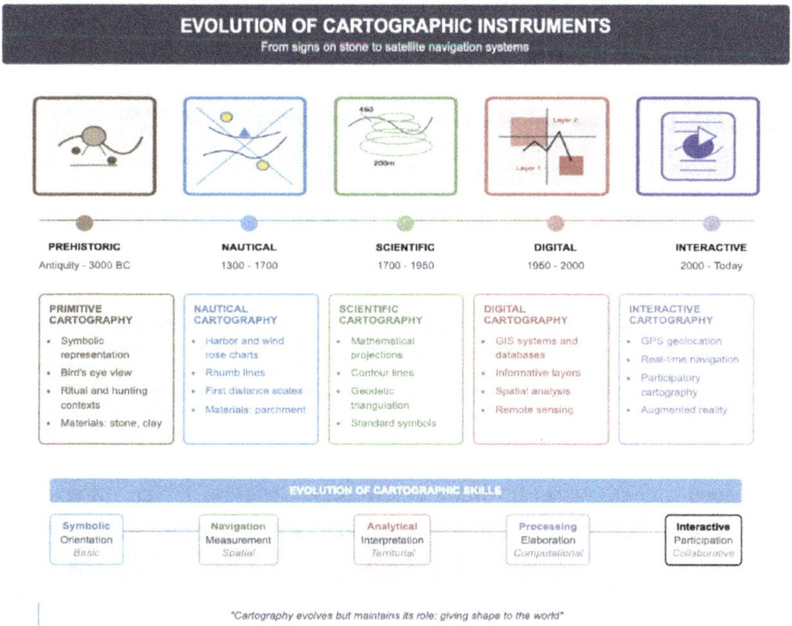

11.2 Cartographic skills in the digital age: educational implications

The enduring value of traditional skills

Despite the proliferation of digital tools, basic map reading skills remain central in education. Orienting oneself in relation to the cardinal points, understanding the relationship between two-dimensional representation and three-dimensional reality, interpreting conventional symbols, and mentally visualizing a landscape from its cartographic representation are skills that transcend technical support and represent essential forms of spatial intelligence.

The ability to "read" the landscape through a map allows us to grasp relationships that escape immediate vision. The morphology of a slope, the hydrographic network, and the development of settlements reveal territorial dynamics that only a trained eye can interpret. These interpretative skills are essential for educating students who are aware of the territorial context in which they live.

Teaching experience shows that these traditional skills are becoming decreasingly common among students. Familiarity with digital technologies does not automatically translate into the ability to read and interpret maps. Many students, accustomed to passively following the instructions of navigation devices, find it difficult to find their bearings on a map, interpret contour lines, and mentally visualize the morphology of the territory represented.

Digital skills and their integration into teaching

Digital cartography requires and develops specific skills that complement traditional ones. The ability to navigate between different levels of information, select relevant data, query geographic databases, and visualize and interpret three-dimensional models represents a new spatial literacy that should be cultivated in education.

Interaction with Geographic Information Systems develops advanced analytical skills. Students can learn to formulate spatial queries, overlay and integrate different types of data, and produce thematic maps that synthesize complex information. These skills are of great value not only in the scientific field, but also for active citizenship.

Geospatial technologies also enable forms of collaboration and sharing that were previously impossible. Participatory cartography, which involves students in mapping their own territory, is a powerful tool for active learning and community empowerment. Projects such as OpenStreetMap can be used in education to develop not only technical skills but also a sense of belonging and responsibility towards the territory.

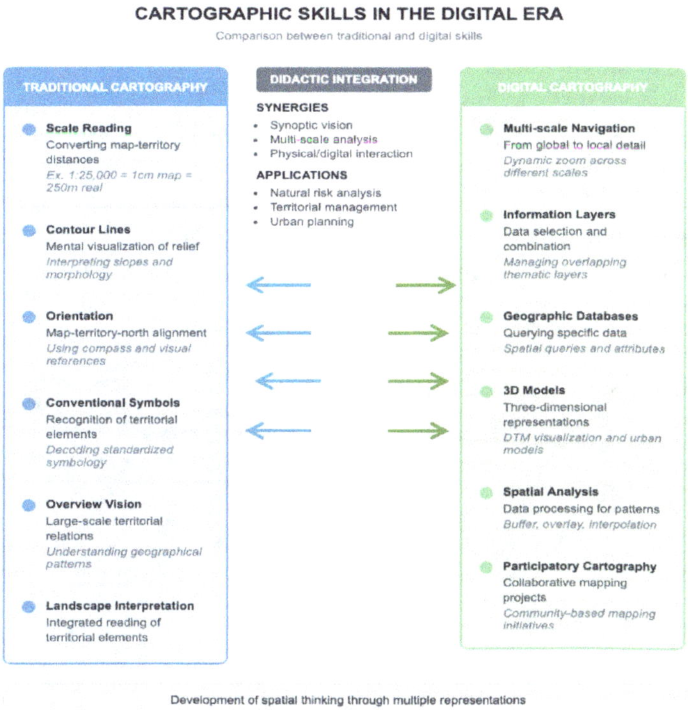

11.3 Cartography as a training tool: experiences and educational perspectives

Integrated training courses: from paper to territory

An effective educational pathway can begin with the analysis of a paper topographic map, which enables students to grasp the overall vision of the region and develop basic skills in orientation and landscape reading. A geological map can then add an interpretative layer, revealing the evolutionary history of the territory and its deep dynamics. The transition to digital systems enriches this knowledge with updated data, three-dimensional visualizations, and specific thematic analyses.

Direct field experience represents the culminating moment of this educational pathway. Physically touching the landscape, recognising on the ground the forms identified on the map, verifying hypotheses formulated through cartographic analysis: these are experiences that consolidate acquired skills and develop a deep awareness of the environment as a complex system.

This integrated methodology allows students to understand the advantages and limitations of each cartographic tool. Traditional maps offer a synoptic vision that facilitates understanding of spatial relationships; digital tools provide updates, interactivity, and advanced analytical capabilities; direct experience adds the sensory and emotional dimension that no representation can replace.

topographic map CTR 2005 | Aerial photo 2012 | Geological map

Different representations of the same area: an example of how topographic, aerial photo, and geological maps provide complementary perspectives on territory (Valle d'Aosta, Italy)

Beyond technique: developing spatial thinking in education

Cartography, whether traditional or digital, is not only a technical tool but also a powerful cognitive mediator that develops specific forms of thinking. Spatial thinking is a fundamental cognitive mode that integrates geographical knowledge, visualization skills, and spatial reasoning. It is the ability to "think spatially," to understand how elements are distributed and interact in space, and to anticipate landscape transformations and dynamics.

Well-designed cartographic activities develop this form of intelligence, stimulating skills that go far beyond the specific technical ability to read a map. Students learn to recognise spatial patterns, integrate different scales of analysis, and understand the relationships between spatial configuration and temporal processes. These skills are transferable to many other disciplines and represent a key form of literacy in the contemporary world.

Spatial thinking is closely linked to systemic thinking, the ability to understand complex systems characterized by multiple interacting elements. The territory is a complex system par excellence, in which physical, biological, and anthropogenic components interact at different temporal and spatial scales. Cartographic analysis, especially when it integrates different layers of information, educates people in this systemic vision, developing a skill that is increasingly in demand in a world characterized by complex and interconnected problems.

11.4 Changing perspective: from user-centrality to understanding the territorial system

A critical aspect of the digital transition concerns the focus of cartography in education. Navigation and geolocation applications literally place the user at the centre of the representation. The blue dot moving across the screen becomes the hub around which the territory is organized, reduced to a backdrop for our personal trajectories.

This user-centric approach undoubtedly has practical advantages, but it risks fuelling an egocentric view of the area that can be problematic from an educational standpoint. Space is no longer perceived as a system with its own internal logic and dynamics, but as a container for our activities. Applications tell us where to go, which roads to take, where to find services and attractions, but they do not necessarily help us understand why the region is organized in a certain way, what processes have shaped the landscape, or what relationships exist between its different components.

In contrast to this egocentric view, cartographic education should promote an ecosystemic perception of the territory, which is not simply a stage for human activities, but a complex system characterized by delicate balances between physical, biological, and anthropogenic components. Understanding these relationships is essential for developing responsible and conscious behaviour in students.

The integrated reading of different cartographic representations can foster this systemic view. A topographic map reveals the morphology of the landscape, a geological map reveals its deep structure, and a land use map shows how humans have modified the natural environment. By integrating these different readings, students can understand how physical characteristics influence human settlements, how human activities transform the landscape, and how natural hazards are often exacerbated by inappropriate interventions.

This integrated reading is particularly significant in regions characterized by geological fragility and intense human modification—whether Mediterranean countries with seismic activity and landslide risk, coastal regions vulnerable to erosion and flooding, mountain areas subject to gravitational processes, or densely populated river basins prone to flooding. Grasping the dynamics of slopes, the regime of watercourses, and the evolution of settlements becomes essential for interpreting phenomena such as landslides, floods, and environmental degradation that affect vulnerable territories worldwide. Cartographic education thus becomes a tool for informed citizenship, educating individuals to understand the risks and potential of the region in which they live.

11.5 Integrated educational courses for cartographic education

The effective integration of traditional and digital cartography requires innovative teaching approaches that enhance the unique features of both media. Below are some examples of successful educational courses that can serve as models for teachers.

Multi-scale explorations of the local territory

A notably effective approach is to analyse the local area using different scales, starting with a detailed topographic map (1:5,000 or 1:10,000) to understand the immediate context, moving on to a medium scale (1:25,000) to grasp broader territorial relationships, and finally to regional scales or beyond to place the study area in its general geographical context.

This multi-scale approach can be enhanced by integrating digital tools such as Google Earth, which allow smooth transitions between different scales. Students can thus understand how certain phenomena are only visible at certain scales, and how a complete knowledge of the territory requires the integration of different perspectives.

Diachronic analysis of territorial changes

Another fruitful educational approach is to analyse the evolution of the morphology over time by comparing historical maps with current representations. The digitization of historical maps and their georeferencing allow precise overlays with contemporary maps, highlighting changes in settlements, modifications to the agricultural landscape, and the evolution of infrastructure.

This type of diachronic analysis develops not only cartographic skills, but also historical awareness and the ability to read the territory as the result of complex temporal processes. The combination of historical paper maps and GIS tools is an excellent example of the integration of traditional and digital media.

Participatory cartography and citizen science

Digital technologies have made forms of participatory cartography previously unthinkable possible. Projects such as OpenStreetMap or citizen science initiatives for environmental monitoring offer valuable opportunities to actively involve students in the creation of geographical knowledge.

These projects can be integrated into educational courses that start with the analysis of official maps to identify aspects that are not adequately represented (minor paths, elements of local interest, microhabitats), which are then mapped by students through field surveys with GPS and smartphones. The information collected can be processed with GIS software and integrated into collaborative platforms.

11.6 Cartography in 21st-century education

Cartography is an educational resource that enhances the development of basic skills for perceiving territorial dynamics. Teaching experiences show that it is possible to effectively integrate traditional and digital tools into educational courses that develop spatial thinking and the ability to critically read the environment.

This integration requires a rethinking of teaching practices that values not only technical skills but also sensitivity to the territory as a complex system. Paper maps, geographic information systems, and digital applications offer complementary perspectives that, when properly integrated, allow for a deep insight of the relationships between physical, biological, and anthropogenic components.

Cartography thus becomes a tool for active citizenship: understanding the region through its representations means acquiring the ability to participate consciously in decision-making processes that affect one's living environment. In an era of rapid environmental change and growing anthropogenic pressures, developing solid cartographic skills remains a fundamental educational goal for training informed and responsible citizens.

TEACHING UNIT: FROM THE MAP TO THE TERRITORY

TITLE: "Reading the landscape through topographic maps"

SCHOOL LEVEL: Middle and high school

DURATION: 3-4 hours (classroom activity) + 4-5 hours (field trip)

OBJECTIVES:

Develop skills in reading and interpreting topographic maps
Understand the relationship between cartographic representation and the real territory
Recognise natural and man-made features of the landscape on paper and on the ground
Orient oneself in space using topographic maps

MATERIALS:

National topographic maps at 1:25,000 scale (or equivalent regional standard)
Compasses
Rulers and string for measuring distances
Structured observation sheets
Coloured pencils for notes
Cameras/smartphones for documentation
(Optional) Tablet with simple GIS apps

PROCEDURE:

Phase 1: Preparation in class (1-2 hours)

Introduction to topographic maps: scale, symbols, contour lines
Guided exercises in reading the map of the area to be visited:
Identification of the main features (settlements, watercourses, relief)
Measuring distances and calculating elevation differences
Tracing the elevation profile along the planned route
Formulating hypotheses about what will be observed during the field trip

Phase 2: Field trip (4-5 hours)

Orienting the map to north and the surrounding landscape

Guided tour with stops at significant points:
Comparing the map and the actual terrain
Identification of features shown on the map
Observation of features not shown or modified compared to the map
Compilation of structured observation sheets
Photographic documentation of observation points

Phase 3: Review in class (1-2 hours)

Comparison of hypotheses formulated, and observations made
Discussion of differences found between the map and the terrain
Reflection on the limitations of cartographic representation
Comparison with digital representations of the area (Google Earth, GIS)

EVALUATION:

Quality of observations and map-territory correlations
Ability to orient oneself using the map
Understanding of the relationships between physical and anthropic elements
Critical reflection on the limitations and potential of different cartographic representations

DIGITAL EXTENSION:

Comparison with the same areas viewed on Google Maps/Earth
Creation of a digital map integrated with photos taken during the outing
Analysis of the differences between traditional paper and digital representation

TEACHING UNIT: PARTICIPATORY DIGITAL CARTOGRAPHY

TITLE: "Mapping our neighbourhood: problems and potential"

SCHOOL LEVEL: Middle and high school

DURATION: 6-8 hours (divided into several sessions)

OBJECTIVES:

Develop digital cartography and GIS skills
Promote critical observation of the local area
Encourage active participation in mapping one's living environment
Stimulate reflection on the critical issues and potential of the area

MATERIALS:

Computers with internet connection
OpenStreetMap account (or similar platform)
Smartphones with geolocation apps
Survey forms
(Optional) Simplified GIS software for schools

PROGRAM:

Phase 1: Introduction to participatory mapping (1-2 hours)

Presentation of the concept of participatory mapping and its uses
Introduction to OpenStreetMap or other collaborative mapping platforms
Guided exercises in basic use of the platforms
Definition of the area to be mapped and elements of interest

Phase 2: Mapping design (1 hour)

Identification of categories of elements to be mapped (e.g., green areas, services, architectural barriers, critical points)
Preparation of structured survey forms
Organisation into working groups and assignment of specific areas
Discussion of the ethical aspects of mapping (privacy, accuracy)

Phase 3: Field survey (2-3 hours)

Exploration of the assigned area with smartphones and survey forms
Geolocation and photographic documentation of the elements identified
Annotation of qualitative observations

Phase 4: Data processing and map creation (2 hours)

Entering the collected data into the mapping platform
Categorizing the mapped elements
Creating a thematic map of the neighbourhood with critical issues and potential
Preparing a presentation of the results

Phase 5: Sharing and reflection (1 hour)

Presenting the results to the class/school/local community
Discussing the issues that emerged and possible solutions
Reflection on the value of participatory mapping as a tool for active citizenship
Evaluation of the experience and skills acquired

EVALUATION:

Quality and accuracy of the mapping work
Ability to critically observe the territory
Collaboration in group work
Skills in the use of digital tools
Quality of the final presentation and reflections

EXTENSION:

Participation in broader citizen science initiatives
Monitoring the evolution of the critical issues identified over time
Comparison with historical maps of the area to analyse changes

TEACHING UNIT: INTEGRATION BETWEEN TRADITIONAL AND DIGITAL CARTOGRAPHY

Title: "From the territory to the map: integrated analysis of hydrogeological risk"

Type: Educational course with a systemic approach and Problem-Based Learning

Target: Secondary school

Duration: 10 hours (classroom, laboratory and field activities)

Approach: Exploratory-investigative, from concrete experience to spatial skills

Learning objectives:

Develop integrated traditional and digital map reading skills
Understand the concepts of hydrogeological risk and territorial vulnerability
Apply spatial thinking to the analysis of environmental issues
Promote awareness of natural risks in the local area

Teaching structure:

Lesson 1: From the territory to risk (2 hours)

Initial phase - "The enigma of the landscape": Observation of aerial photos of the local area with the guiding question: "What elements of this territory could be dangerous?"
Investigative phase: Introduction to the concepts of hazard, vulnerability, and risk through local case studies
Exploratory phase: Initial reading of topographic maps to identify significant morphological features

Lesson 2: The map tells the story (3 hours)

Practical workshop: Integrated reading of topographic, geological, and land use maps
Analysis activity: Identification of the hydrographic network, analysis of slopes, identification of settlements at risk

Digital experiment: Introduction to GIS with georeferencing and overlaying of information layers

Lesson 3: Time in the area (3 hours)

Historical investigation: Research into past events and their location on maps
Diachronic analysis: Comparison of historical and recent satellite images
Modelling: Creation of derived maps (slopes, exposures) with GIS tools

Lesson 4: From knowledge to action (2 hours)

Integrated synthesis: Development of a simplified local risk map
Presentation: Each group presents its analysis of its territorial sector
Discussion: Proposals for interventions and mitigation strategies

Skills developed: Integrated reading of cartographic representations, spatial and systemic thinking, understanding of territorial dynamics, awareness of environmental risks, scientific communication.

Interdisciplinary links: Geography (territory and transformations), Science (geomorphological processes), History (events and local memory), Mathematics (quantitative analysis), Civic education (prevention and civil protection).

Final product: Digital atlas of local hydrogeological risk with proposals for action, to be shared with the school community and local authorities.

ASSESSMENT GRID FOR CARTOGRAPHIC ACTIVITIES

Competency	Basic Level	Intermediate Level	Advanced Level
Map Reading	Recognizes main symbols and identifies evident elements	Correctly interprets symbols and conventions, understands morphology	Performs integrated and in-depth reading, grasps complex relationships between diverse elements
Orientation	Orients with the map in simple situations	Correctly uses compass and orients the map with respect to territory	Orients confidently in complex contexts integrating map, compass and territory observation
Use of Digital Tools	Uses basic functions of digital maps and GIS	Navigates between different information layers and queries geographic databases	Develops complex spatial analyses and integrates data of diverse nature
Spatial Thinking	Understands simple spatial relationships	Analyzes spatial patterns and relationships at different scales	Develops complex interpretative models of territorial dynamics
Representation Capacity	Produces simple cartographic sketches	Develops coherent and readable thematic maps	Creates innovative and effective cartographic representations to communicate complex phenomena
Systemic Vision	Recognizes some connections between territory elements	Understands main flows and cycles that characterize the territorial system	Analyzes territory as a complex system, identifying feedback and evolutionary dynamics

Capelinhos Volcano, the most recent in the Azores_ Faial Island

Further reading

Bednarz, S.W., & Kemp, K. (2011). *Understanding and nurturing spatial literacy*. Procedia - Social and Behavioural Sciences, 21, 18-23. DOI: https://doi.org/10.1016/j.sbspro.2011.07.004

Goodchild, M.F., & Janelle, D.G. (2010). *Toward critical spatial thinking in the social sciences and humanities*. GeoJournal, 75(1), 3-13. DOI: https://doi.org /10.1007/s10708-010-9340-3

Lee, J., & Bednarz, R. (2012). Components of spatial thinking: Evidence from a spatial thinking ability test. Journal of Geography, 111(1), 15-26. DOI: https://doi.org/10.1080/00221341.2011.583262

Haklay, M. (2016). *Citizen Science and Volunteered Geographic Information: Overview and Typology of Participation*. In: Capineri, C. et al. (eds.) *European*

Handbook of Crowdsourced Geographic Information. Ubiquity Press, London, pp. 105-122.

Brus, J., Voženílek, V., & Popelka, S. (2013). *Modern Education in Thematic Cartography*. Palacký University Olomouc. https://www.researchgate.net/publication/259583080_Modern_Education_in_Thematic_Cartography

Kerski, J.J. (2015). "Geo-awareness, Geo-enablement, Geotechnologies, Citizen Science, and Storytelling: Geography on the World Stage." *Geography Compass*, 9(1), 14-26. https://doi.org/10.1111/gec3.12193

Montello, D.R. (2009). "Cognitive Research in GIScience: Recent Achievements and Future Prospects." *Geography Compass*, 3(5), 1824-1840. https://doi.org/10.1111/j.1749-8198.2009.00273.x

Schultz, R.B., Kerski, J.J., & Patterson, T.C. (2008). "The Use of Virtual Globes as a Spatial Teaching Tool with Suggestions for Metadata Standards." Journal of Geography, 107(1), 27-34. https://doi.org/10.1080/00221340802049844

CHAPTER 12.
USING THE TERRITORY AS A TEACHING TOOL: GEOSITES AS LEARNING OBJECTS

Field trips are an indispensable teaching tool for any scientific and naturalistic discipline. Direct experience in the environment is a fundamental complement to theoretical learning, capable of transforming abstract concepts into tangible, observable, and measurable realities. No textbook, digital simulation, or classroom explanation can replace the educational value of direct experience in the field, where the complexity of the real-world manifests itself in its entirety.

As a learning environment, territory offers extraordinary educational opportunities that go far beyond the acquisition of specific disciplinary knowledge. Fieldwork stimulates the development of cross-curricular skills: teamwork, where each student takes on roles and responsibilities; systematic observation and problem-solving skills; adaptation to unexpected situations; respect for safety rules and the natural environment. These skills are impossible to develop fully within the school walls.

Educator should emphasize that a trip to an area of environmental value should not be seen as a simple "outing" or recreational experience, but as an authentic educational tool with specific learning objectives and rigorous methodologies. The difference lies not in the destination, but in the approach: an educational field experience requires preparation, methodology, investigative tools, and opportunities to reflect on the experiences.

In this educational context, geosites and geoparks emerge as extraordinary "open-air laboratories" particularly suited to the study of geosciences.

Geosites—defined in international geological heritage initiatives and catalogued by national geological surveys (such as ISPRA in Italy at https://www.isprambiente.gov.it/geosites, the British Geological Survey, USGS in the United States, and similar institutions worldwide) represent the geodiversity of a territory, understood as the range of geological, geomorphological, hydrological, and pedological characteristics of a given area. Given that these characteristics are decisive for the various species living in these territories, it can be considered that the conservation of geodiversity and the protection of geological heritage contribute to combating biodiversity loss and maintaining the integrity of ecosystems. According to the commonly accepted definition, "*a geosite can be defined as a location, area or territory where it is possible to identify a geological or geomorphological interest for conservation* (W.A. Wimbledon, 1996)."

These places, characterized by geological features of particular scientific, aesthetic, and educational value, offer the opportunity to understand complex natural processes through direct observation of their manifestations in the region. From fault systems to stratigraphic sequences, from landscape forms to erosion phenomena, geosites make visible and understandable processes that generally develop on temporal and spatial scales that transcend everyday experience.

12.1 The territory as a learning environment: requirements for an effective approach

The use of the territory as a learning environment requires specific characteristics that transform the experience from passive to active, making the geosite an authentic teaching tool.

Active participation and direct investigation are the first basic requirements. Students cannot be mere observers but must become researchers who conduct field investigations, collect data, analyse samples, and formulate hypotheses. This active dimension is essential for developing authentic scientific skills and a meaningful relationship with the landscape. When students feel that they are protagonists in the investigation, their motivation and depth of learning increase significantly.

The **careful selection of learning objects** is the second crucial element. Not all geosites are equally suitable for educational purposes; their selection requires an assessment that simultaneously considers scientific value, physical and intellectual accessibility, and the possibility of direct observation of geological phenomena. A site that is geologically significant but conceptually too complex for a given age group, or physically difficult to reach, may not be the best choice. Educators must balance these factors to identify sites that offer the best learning opportunities for their specific students.

A **structured investigative approach** transforms simple observation into true scientific inquiry. The field experience should be guided by "driving questions" that stimulate students' curiosity and scientific reasoning. These questions guide observation and data collection, transforming the territory into a problem to be solved, a text to be decoded. The questions should be open-ended enough to stimulate exploration, but also specific enough to focus students' attention on the most significant elements.

Intentional multidisciplinary enriches the learning experience. Teachers must be able to highlight not only geological aspects, but also connections with physics, chemistry, biology, history, and local culture. For example, an outcrop of carbonate rocks can become an opportunity to explore the chemical processes of dissolution, the ecology of the organisms that contributed to the formation of the rock, and the historical use of those materials in local architecture. This integrated approach fosters in the students the be aware of the environment as a complex system and the developing a multidimensional view.

Unexpected connections are a powerful teaching tool. When students discover, for example, that a particular geological formation has influenced the urban development of a city, or that an ancient earthquake led to the abandonment of a historic settlement, these 'eureka moments' are particularly significant from a learning perspective.

Adequate teaching time is essential for meaningful learning. Careful observation, reflection, discussion, and consolidation require sufficient time. Overly compressed field experiences reduce students to mere tourists, while extended time allows for the cognitive processing necessary to transform the experience into lasting learning. Time is a fundamental variable in developing a meaningful relationship with the territory.

The **specific preparation of educators** is perhaps the most critical requirement. Teachers must possess not only solid geological knowledge but also specific pedagogical skills to transform geological elements into effective teaching tools. This dual competence is essential to effectively mediate between the complexity of geological phenomena and the cognitive abilities of students. Educators must be able to 'translate' geological complexity into accessible experiences, without trivializing but making even the most complex concepts understandable.

	Primary School	Lower Secondary School	Upper Secondary School - First Biennium	Upper Secondary School - Second Biennium
Example of vertical progression of competencies:	Sensory exploration of local territory, recognition of basic forms and materials, first classifications of rocks and landscapes.	Simple mapping of geological elements, recognition of main rock types and their characteristics, understanding of observable erosive and sedimentary processes.	Quantitative analysis of physical and chemical parameters, reconstruction of local geological history, assessment of simple geological risks.	Structured geological investigation with data collection and analysis, correlation between local phenomena and global geological processes, assessment of anthropic impact on the geological system.

To make effective use of the landscape for educational purposes, it is necessary to implement specific strategies that involve all stages of the educational experience, from preparation to follow-up.

Pre-field preparation: building cognitive foundations

The preparatory phase is essential for focusing students' attention and providing the necessary interpretative keys:

Pre-field preparation includes a preliminary conceptual introduction in the classroom, the formulation of hypotheses to be tested in the field, the preparation of observation and data collection tools, and the discussion of prior knowledge and possible misconceptions. This phase helps students to develop an informed perspective, capable of going beyond the visible surface of the territory.

Field investigation methodologies: from territory to knowledge

At the heart of the teaching approach are field investigation methodologies, which transform direct experience into structured learning:

Field investigation uses structured observation sheets that guide students' attention to significant elements, measurement and sampling tools that allow quantitative data to be collected, systematic documentation through drawings, photographs, and notes that record observations, and the collection of quantitative and qualitative data that allow for subsequent analysis.

Students learn to observe, measure, document, formulate hypotheses, and test them, developing scientific skills that go far beyond the simple memorization of information. The region becomes an open book that students learn to read and interpret.

Post-fieldwork processing: consolidating learning

The processing phase is essential for consolidating learning and transforming experience into structured knowledge and a tool for evaluating learning and the skills that may have been acquired.

Post-field processing includes analysing the data collected, comparing observations with initial hypotheses, developing explanatory models that give meaning to the observations, presenting and discussing the results to share the knowledge acquired, and connecting with other geological contexts that broaden the perspective beyond the single site visited.

This phase allows for the systematization of knowledge, the linking of direct experience to theoretical concepts, and the development of metacognitive skills through reflection on one's own learning process.

THE DIDACTIC CYCLE OF TERRITORY FIELD EXPERIENCE

PRE-FIELD PHASE	FIELD PHASE	POST-FIELD PHASE
Classroom Preparation	Geosite Exploration	Processing and Synthesis
ACTIVITIES	ACTIVITIES	ACTIVITIES
• Conceptual introduction of elements • Formulation of hypotheses to verify • Study of satellite images • Identification of initial preconceptions • Preparation of necessary tools • Layout of survey sheets • Discussion of safety rules • Definition of group roles • Research of previous information	• Structured observation with sheets • Sample collection (if authorized) • Measurement of physical parameters • Photographic documentation • Field sketching • Comparison of group observations • Preliminary data analysis • Identification of field relationships • Discussion with guide/teacher	• Analysis of collected data • Verification of initial hypotheses • Development of explanatory models • Presentation of results • Connection with theoretical concepts • Comparison with other geological contexts • Preparation of communication materials • Metacognitive reflection • Learning assessment
OBJECTIVES	OBJECTIVES	OBJECTIVES
• Provide interpretative keys • Orient student attention • Activate preconceptions and motivation • Develop predictive abilities • Encourage research questions	• Develop scientific methodology • Collect meaningful data • Promote direct investigation • Stimulate observational skills • Encourage peer cooperation	• Systematize knowledge • Develop explanatory models • Transfer skills to other contexts • Develop metacognition • Consolidate learning

The cycle repeats with increasing levels of complexity and depth →

Use of interdisciplinary links: creating meaning

A distinctive feature of the teaching approach is its ability to highlight unexpected connections between geological phenomena and other fields, creating moments of meaningful learning:

These connections may relate to historical impact (as in the case of landslides that have altered communication routes), influence on microclimates and ecosystems, relationships with past and present human activities, and implications for future land-use planning. These cross-cutting connections allow students to understand the territory as a complex system and to develop integrated thinking.

When students discover, for example, that the distribution of settlements in a valley is determined by the geology of the substrate, or that a particular microclimate depends on the geological conformation, cross-cutting links are created that make learning particularly meaningful.

12.2 Curriculum integration and logistical challenges

The effectiveness of the educational approach to the territory requires full integration into the school curricula. It is not a question of occasionally adding a field trip, but of rethinking the educational path to include experiences outside in an organic way. This happens through organic inclusion in the curriculum, conceiving field experiences not as occasional or ancillary activities, but as regular and central components of the educational path. Continuity between field experience and classroom learning ensures that experiences in the local area do not remain isolated but are intertwined with the theoretical concepts developed in the classroom. The assessment of learning acquired through local experience formally recognises the educational value of these experiences, integrating them into the assessment system. The vertical progression of experiences through the different school cycles facilitates the development of increasingly complex skills, with local experiences evolving in complexity and depth as the school career progresses.

This integration requires a rethinking of school curricula, which should explicitly include local activities as regular components of the educational path. It also requires appropriate assessment systems that can capture the skills developed through these direct experiences.

The logistical requirements of the educational approach to the local area present specific challenges that cannot be ignored. The extended time required for observation and investigation is difficult to reconcile with the traditional school timetable. A meaningful experience in the local area requires extended periods of time, which are incompatible with one- or two-hour lessons. Specific equipment for data collection is a necessary but often problematic investment for schools with limited resources. Geological compasses, geologist's hammers, magnifying glasses, measuring instruments, sample analysis kits: these are tools that greatly enrich the educational experience but require dedicated resources. Small groups are essential to ensure the active participation of all students. Unlike a classroom lecture, fieldwork requires a much more favourable teacher-student ratio, allowing for meaningful interaction and adequate supervision. The need to access lesser known but scientifically significant sites requires in-depth knowledge of the local area and, often, the opening of non-tourist routes. The most famous and accessible sites are not necessarily the most significant from an educational point of view.

These logistical requirements often represent a practical obstacle to the systematic implementation of the educational approach to the local area. It is necessary to rethink traditional school organisation, overcoming the rigidity of timetables and spaces, and to develop partnerships with local authorities, parks, and natural history museums that can provide logistical and scientific support.

12.3 Digital technologies as a tool for enhancing the field experience

The integration of digital technologies into the field experience should not replace direct observation but can significantly enhance it. Geolocation apps allow observation points to be recorded accurately, facilitating the creation of thematic maps and data sharing. Georeferenced photography allows the elements observed to be systematically documented, automatically linking them to their geographical location.

Simple mobile GIS systems allow observations to be superimposed in real time on existing thematic maps, facilitating integrated analysis. Portable sensors connected to smartphones allow environmental measurements (pH, temperature, humidity) to be taken, enriching qualitative analysis with quantitative data.

Online collaborative platforms allow observations to be shared immediately, promoting teamwork even during field experience.

When used with methodological awareness, these technologies do not distract attention from outdoor experiences but amplify the capacity for observation, documentation, and analysis. They also allow continuity between field experience and subsequent classroom work, facilitating data processing and the sharing of results.

12.4 Geosites as learning objects

The educational use of the environment is a powerful approach to teaching geosciences. From this perspective, geosites are not simply "places to visit," but authentic "learning objects" that enable a transformative learning experience.

This approach can promote a widespread culture of geological complexity, the interdependence between natural phenomena and human activities, and the importance of geoconservation. Through well-structured educational field experiences, students can develop not only geological knowledge but also investigative skills, integrated analysis abilities, and environmental awareness, which are fundamental elements for facing the environmental challenges of the future. Then, it becomes not only an object of study but also a tool for educating informed citizens who are able to critically interpret their environment and actively participate in its sustainable management.

In this perspective, the educational valorisation of geosites represents a strategic investment not only to improve the teaching of geosciences, but also to promote a new culture of the region based on knowledge, respect, and responsibility. It is an approach that recognises the educational value of geological heritage and transforms it into a resource for the formation of essential skills in the contemporary world.

Experience shows that when students can interact directly with the territory through appropriate teaching methods, they develop not only more solid and lasting knowledge, but also an emotional connection with the environment that is necessary to motivating responsible behaviour. From an indistinct backdrop to our daily activities, it becomes a text rich in meaning that they learn to decipher, understand, and respect.

Regional Examples as Methodological Templates

The following three educational formats illustrate how specific geological features can be transformed into effective learning experiences. While these examples draw from sites in the Italian Alps—a region offering exceptional geological diversity and accessibility—the pedagogical frameworks they demonstrate are universally applicable. Teachers worldwide can adapt these structures to their own regional geological contexts: glacial deposits exist in formerly glaciated regions across North America, northern Europe, Patagonia, and New Zealand; landslide phenomena occur in mountain regions globally; and mining sites reflecting tectonic processes can be found on all continents.

The value of these examples lies not in their specific geographical location but in demonstrating how to transform a geological feature into a "learning object"—a structured, purposeful educational resource that facilitates specific learning outcomes. Each format shows how to progress from initial observation through systematic investigation to synthesis and interpretation, treating the geological site as an active teaching tool rather than a passive destination. This approach creates a pedagogical framework transferable to analogous geological features in any region. Teachers should view these as templates for developing similar activities based on their local geological heritage, maintaining the investigative structure and learning object design principles while substituting regionally appropriate content.

TEACHING UNIT: Fluvial-glacial deposits as a climate archive

Title:" Reading the climate of the past: sediments tell the story"

Type: Exploratory educational course with a paleoclimatic approach

Target: Secondary school

Duration: 8 hours (preparation, fieldwork, laboratory)

Approach: Investigative-interpretative, from the specific to the general

Learning objectives:

Understand the concept of geological time through sediments
Develop stratigraphic reading and paleoclimatic interpretation skills
Link past and present climate change
Recognise the value of sediments as an environmental archive

Educational structure:

Phase 1: Introduction to climate time (1 hour)

Guiding question: "How do we know what the climate was like before thermometers existed?"
Conceptual exploration: Discussion of climate proxies and natural archives
Theoretical preparation: Recognition of different types of glacial sediments

Phase 2: Stratigraphic reading in the field (4 hours)

Layer identification: Mapping and description of different sedimentary horizons
Systematic sampling: Collection of representative samples from each level
Measurement of parameters: Grain size, thickness, sedimentary structures
Environmental correlation: Linking sedimentary characteristics and climatic conditions

Phase 3: Paleoclimatic laboratory (3 hours)

Sediment analysis: Grain size and mineralogical study of samples

Chronological reconstruction: Construction of the sequence of climate oscillations

Present-past comparison: Discussion of contemporary climate change

Future modelling: Projections based on natural climate cycles

Skills developed: Stratigraphic reading, inferential thinking, understanding of geological time, paleoclimatic analysis, systemic view of the Earth's climate

Interdisciplinary links: Geology (stratigraphy), Chemistry (sediment analysis), Physics (climate cycles), Geography (paleoclimates), Civic education (climate change).

Final product: Local climate archive with a chronology of past oscillations and comparison with current climate trends.

Fluvioglacial deposits: horizons with different grain sizes

TEACHING UNIT: A river barrier landslide as a geological laboratory

Title: "Territorial detectives: investigation of a historic landslide"

Type: Investigative educational course with a systemic approach

Target: Middle and high school

Duration: 6 hours (preparation, fieldwork, processing)

Approach: Problem-Based Learning from direct observation to historical reconstruction

Learning objectives:

Develop observation and spatial representation skills of the territory
Understand the dynamic processes and causal sequences of landslide phenomena
Integrate geology, local history, and risk assessment
Reconstruct past events through traces in the landscape

Teaching structure:

Phase 1: Investigative preparation (1 hour)

Initial puzzle: "Why shouldn't this lake exist?" Observation of aerial photos of the area
Formulation of hypotheses: Discussion of possible causes of the formation of the basin
Preparation of tools: Survey sheets, topographic maps, measuring instruments

Phase 2: Field investigation (3 hours)

Detective mapping: Identification and delimitation of the detachment zone, landslide body, and accumulation zone
Collecting clues: Study of the rocks involved, measurements, photographic documentation
Chronological reconstruction: Analysis of traces to reconstruct the sequence of events

Phase 3: Analysis workshop (2 hours)

Data processing: Analysis of the samples collected, and measurements taken
3D reconstruction: Creation of models of the situation before and after the landslide
Risk assessment: Simulation of future scenarios and impacts on settlements and infrastructure
Presentation: Each group presents its reconstruction of the event

Skills developed: Systematic observation, territorial mapping, causal thinking, historical reconstruction, risk assessment, scientific communication.

Interdisciplinary links: Geology (gravitational processes), Physics (fluid mechanics), History (local events), Geography (territorial impact), Technology (monitoring systems).

Final product: Complete investigative dossier on the landslide with chronological reconstruction, risk analysis, and monitoring proposal.

The Mont Avi-Montjovet landslide (AO)

TEACHING UNIT: A Mining Site as a Window into Global Tectonics

Title: "From Mines to Plate Tectonics: Traces of Global Processes"

Type: Interpretative educational pathway with multiscale approach

Target: Upper secondary school

Duration: 10 hours (preparation, field, laboratory, processing)

Approach: Inductive from local to global scale

Learning Objectives:

Understand the relationship between mineral deposits and tectonic processes
Develop correlation skills between local phenomena and global dynamics
Interpret geological history through minerals and structures
Connect economic geology and plate tectonics

Educational Structure:

Phase 1: The Mystery of Metals (2 hours)

Guiding question: "Why were these minerals found exactly here?"
Initial exploration: Analysis of the geographical distribution of similar deposits worldwide
Theoretical preparation: Introduction to mineral deposit formation processes

Phase 2: Reading the Mining Site (4 hours)

Geological mapping: Identification of host rocks and tectonic structures
Mineralogical study: Recognition of minerals and paragenetic associations
Structural analysis: Measurement of faults, fractures and mineralization orientation
Targeted sampling: Collection of representative samples for analysis

Phase 3: Geochemical Laboratory (3 hours)

Mineralogical analysis: Study of samples with hand lens and stereomicroscope
Geochemical interpretation: Correlation between composition and formation environment

Genetic modelling: Hypotheses on deposit formation processes

Phase 4: From Mine to World (1 hour)

Global correlation: Comparison with similar deposits on other continents
Paleogeographic reconstruction: Positioning the site in the context of continental drift
Evolutionary synthesis: Integration of local history into global tectonics
Presentation: Communication of results and identified correlations

Developed Competencies: Mineralogical analysis, structural interpretation, multiscale thinking, spatial -temporal correlation, understanding of global tectonic processes.

Interdisciplinary Connections: Geology (mineralogy and tectonics), Chemistry (geochemical processes), Physics (thermodynamics), Geography (global distribution), History (local extractive activities).

Final Product: Interpretative model of the deposit with reconstruction of its genesis in the context of regional tectonic evolution and correlations with global plate dynamics.

Copper deposits in the Hérin mines – Champdepraz

GEOSITE OBSERVATION SHEET

Structured tool for field investigation

1. GEOSITE IDENTIFICATION DATA

Geosite name:

Geographic location: GPS Coordinates Altitude (m a.s.l.)

Survey date: Weather conditions:

Working group:

2. GEOLOGICAL CHARACTERISTICS

Main typology:

Geomorphological
Stratigraphic
Hydrogeological
Tectonic/Structural
Paleontological
Mineralogical/Petrographic
Other:

Lithological description:

Describe the rock types, their characteristics, relationships and observable structures...

3. OBSERVATIONS AND MEASUREMENTS

Observable morphological elements:

Describe landforms, erosional features, depositional structures, etc.

Measurements and sampling:

Parameter	Measured Value	Instrument Used

4. INTERPRETATION AND CONNECTIONS

Preliminary interpretation of observed processes:

Describe the geological processes that may have formed or modified this site...

Connections with other systems (biological, anthropic, historical):

Identify relationships with ecosystems, human activities, local history...

SPACE FOR SKETCHES AND VISUAL DOCUMENTATION

Hemis High Altitude National Park, Ladakh, India

Further reading

Reynard, E., & Brilha, J. (Eds.) (2018). *Geoheritage: Assessment, Protection, and Management*. Elsevier.

Panizza, M. (2001). "Geomorphosites: concepts, methods and examples of geomorphological survey." *Chinese Science Bulletin*, 46(1), 4-6.

Henriques, M.H., Pena dos Reis, R., Brilha, J., & Mota, T. (2011). "Geoconservation as an Emerging Geoscience." *Geoheritage*, 3(2), 117-128.

Newsome, D., & Dowling, R. (Eds.) (2010). Geotourism: The Tourism of Geology and Landscape. Goodfellow Publishers.

Zouros, N. (2004). "The European Geoparks Network: Geological heritage protection and local development." Episodes, 27(3), 165-171.

Farsani, N.T., Coelho, C., & Costa, C. (2011). "Geotourism and geoparks as novel strategies for socio-economic development in rural areas." *International Journal of Tourism Research*, 13(1), 68-81.

Popa, M.E., Kędzior, A., & Fodolică, V. (2010). "The Geopark as a vehicle for environmental education: The case of the Hateg Country Dinosaurs Geopark, Romania." Studia UBB Geologia, 55(1), 71-74.

Occhipinti, S. (2021). Geoparks and Geo-Sites: Geological "Learning Objects". In: Singh, R., Wei, D., Anand, S. (eds) Global Geographical Heritage, Geoparks and Geotourism. Advances in Geographical and Environmental Sciences. Springer, Singapore. https://doi.org/10.1007/978-981-15-4956-4_12

Diolaiuti, G., & Smiraglia, C. (2010). "Changing glaciers in a changing climate: how vanishing geomorphosites have been driving deep changes in mountain landscapes and environments." *Géomorphologie: relief, processus, environnement*, 16(2), 131-152.

Coratza, P., & Panizza, M. (Eds.) (2010). *Geomorphology and Cultural Heritage*. Memorie Descrittive della Carta Geologica d'Italia, 87, ISPRA, Rome.

CHAPTER 13.
THE DATA MINING TEST:
AN INNOVATIVE TEACHING TOOL FOR GEOSCIENCES

The Data Mining Test (DMT) is an innovative teaching approach that effectively responds to the challenges of teaching Earth Sciences, a discipline that often suffers from a limited perception despite its extraordinary educational potential. The pandemic and the increase in distance learning have accelerated the need to rethink assessment tools, pushing towards methodologies that go beyond the traditional frontal approach and multiple-choice quizzes, which are particularly ineffective when students are represented only by letters or symbols on a dark screen.

DMT derives from the concept of data mining developed in the business world in the late 1980s, defined as the set of techniques and methodologies aimed at extracting useful information from large amounts of data. In the educational context, adapted to education, this approach creates a powerful tool assessing knowledge while developing fundamental cross-cutting skills.

13.1 Data selection and organisation

Building an effective Data Mining Test requires particular attention during the data selection and organisation phase. A huge amount of information is available on the web, but as described in the specialist literature, we often find ourselves in a situation that is "rich in data but poor in information," and therefore in knowledge. It is essential to transform this abundance of raw data into meaningful educational resources.

For the search for appropriate data, there are numerous national and international portals that offer access to structured datasets related to geological and environmental phenomena

Data Resources for DTM Construction in Earth Sciences

ITALIAN DATABASES	
Database	**URL and Description**
ISPRA - National Geoportal	https://www.isprambiente.gov.it/it/banche-dati Geological maps, risk maps and environmental databases
INGV	http://terremoti.ingv.it/opendata Seismic and volcanic data, time series and interactive maps
SNPA	https://www.snpambiente.it/dati/ Environmental data comparable across different regions
Italian Geological Service	https://www.isprambiente.gov.it/it/servizi/servizio-geologico-ditalia Digital geological maps and thematic databases
Regional Geoportals	Example: http://www.geoportale.piemonte.it Regional technical maps and hydrogeological planning

FRENCH AND EUROPEAN DATABASES	
Database	**URL and Description**
EduMed	https://edumed.unice.fr/ Geological and seismological data of the Mediterranean area with educational resources
BRGM - Géoportail	https://www.geoportail.gouv.fr/ French cartographic and geological portal
InfoTerre	http://infoterre.brgm.fr/ Detailed geological data of France
Géorisques	https://www.georisques.gouv.fr/ Data on natural and technological risks in France
OneGeology Europe	http://www.onegeology-europe.org/ Integrated pan-European geological portal

INTERNATIONAL DATABASES	
Database	**URL and Description**
USGS	https://www.usgs.gov/products/data Wide range of geological and geophysical data at global level
NASA Earth Data	https://earthdata.nasa.gov/ Global satellite observations and climate data
NOAA	https://www.ncdc.noaa.gov/cdo-web/ Climate and oceanographic data with time series
Copernicus Climate Change	https://climate.copernicus.eu/data-access European satellite data and climate models

When selecting data sources for a DMT, it is advisable to look for data that is easily accessible and downloadable, preferably in standard formats that can be processed with common tools, to favour datasets allowing a phenomenon to be examined from different perspectives, integrating quantitative and qualitative data, select case studies presenting well-documented geological and environmental problems, and verify the quality and reliability of the data, preferring recognised institutional or scientific sources.

Effective DMT requires data with certain key characteristics:

- Accessibility: data must be easily accessible t, without requiring special credentials or proprietary software. It is useful to check the accessibility of resources in advance and, if necessary, provide directly some materials that may be difficult to download.
- Variety of formats: it is advisable to integrate different types of data (images, maps, graphs, tables, time series) to stimulate different analytical skills and encourage a multidimensional approach to the problem.
- Interconnection: the selected data should allow connections to be established between different aspects of the phenomenon being studied, facilitating a systemic approach and recognising cause-effect relationships.
- Gradual complexity: the dataset should allow questions of varying difficulty to be structured, from the simplest to the most complex, adaptable to different school levels and different student skills.

13.2 The methodological structure of DMT

The DMT divides the Earth Sciences teaching-learning process into several phases that reflect the skills to be developed:

The first phase consists of identifying relevant information within datasets, maps, graphs, or tables. Students must learn to navigate between different data sources and select those relevant to the question asked.

This is followed by the analysis of the case, problem, or phenomenon under examination, which requires the ability to understand and contextualize the data within the framework of basic Earth Science knowledge.

They must then develop strategies to identify useful or required data, distinguishing it from irrelevant or redundant data, a crucial skill in the age of information overload.

The next step involves identifying links and relationships between different pieces of information, requiring synthesis and complex thinking skills, which are specifically relevant in geosciences, which examine complex and interconnected systems.

Finally, students must visualize and communicate their findings using appropriate tools, developing effective scientific communication skills.

For effective implementation, it is recommended to limit the duration of the test to approximately 45 minutes to maintain student attention and engagement, to use a limited number of documents, links, and images (no more than 6) to avoid cognitive overload, and finally to include a maximum of 20 questions, which can be either open-ended or closed-ended, structured in such a way as to guide the analysis progressively.

But DMT does not start from scratch: students must have some basic knowledge of Earth Sciences to effectively analyse data related to specific phenomena. As highlighted in educational research, "it is not possible to analyse data related to a specific phenomenon if one does not know the phenomenon in general."

From Data to Decision: DTM Cognitive Pathway

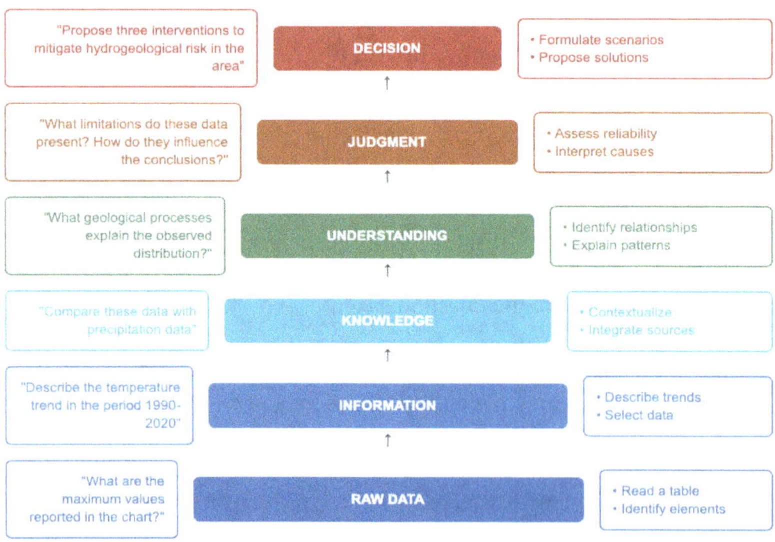

The added value of DMT lies in its ability to simultaneously assess both hard skills (specific technical skills of the discipline) and soft skills (transversal skills such as critical thinking, problem solving, and digital literacy). In addition, DMT promotes a metacognitive and reflective learning process, requiring students not only to find answers, but to understand the processes through which these answers are developed.

13.3 The educational value of the Data Mining Test

Data Mining Tests innovatively transform geoscience teaching, transforming learning from a passive process of memorization into an active experience of knowledge construction.

One of the most significant potentials of DMT is its ability to integrate theoretical knowledge and practical skills. Students do not learn abstract concepts but apply them directly to the analysis of real cases, developing what is called "usable knowledge." The geology of glaciers or alpine slopes becomes a tool for interpreting concrete phenomena and understanding current environmental issues.

DMT also promotes an integrated knowledge of geological phenomena, overcoming the fragmentation typical of traditional teaching. They are guided to recognise the interconnections between different components of the Earth system: how geology influences morphology, how this conditions the distribution of vegetation, how these factors together determine the distribution of natural hazards.

Another added value of DMT is the simultaneous development of disciplinary and cross-cutting skills. While acquiring specific knowledge about geology, they also hone skills such as critical thinking, data analysis, problem solving, and scientific communication.

The most innovative aspect of DMT is its ability to develop 'territorial literacy' in students, i.e. the ability to read and interpret the landscape in its physical and anthropic components, to recognise potential risks, and to formulate hypotheses about possible future developments. This skill is essential for training informed citizens who can actively participate in decision-making processes concerning land management.

The Data Mining Test thus represents a paradigm shift in the teaching of Earth Sciences: from a content-centred approach to a skills-centred approach, from a fragmented view to an integrated view, from passive learning to active and meaningful learning.

Types of Questions in Data Mining Tests

Type	Function	Application Example
Descriptive Questions	Identification and description of patterns in data	*"Describe the trend in average temperatures for the period 1990-2020"*
Comparative Questions	Comparison between different datasets or portions of the same dataset	*"Compare the precipitation trend in the two valleys"*
Interpretative Questions	Explanation of observed patterns	*"What geological processes can explain the observed distribution of landslide phenomena?"*
Predictive Questions	Formulation of predictions based on identified trends	*"Assuming the glacial retreat rate remains constant, what will be the front position in 2030?"*
Evaluative Questions	Critical assessment and reliability of data	*"What limitations do the analyzed data present? How might they influence the conclusions?"*
Creative Questions	Development of solutions or alternative scenarios	*"Propose three possible interventions to mitigate hydrogeological risk in the area"*

Questions should be structured progressively, from the simplest to the most complex
An effective DMT includes questions of different types, adapted to the student level

13.4 DMT and active teaching methods: synergistic integrations

The Data Mining Test can be effectively integrated with other active teaching methods, creating richer and more multidimensional learning paths:

Main integrations:

- **Problem-Based Learning (PBL):** DMT provides the tools to analyse the data necessary to understand real-world problems before proceeding to develop solutions.
- **Service Learning:** Data analysis can be used for projects with a concrete impact on the local community, such as the creation of risk maps or information materials for civil protection.
- **Flipped Classroom:** The theoretical part is studied independently at home, while class time is devoted to collaborative data analysis under the guidance of the teacher.

- **Inquiry-Based Learning:** DMT is structured as a progressive investigation, where students formulate initial hypotheses and test them through the analysis of additional data.
- **Project-Based Learning:** Data analysis is integrated into broader, interdisciplinary projects, such as studies on territorial sustainability.
- **Outdoor Education:** Data analysis in the classroom can be preceded or followed by field trips to collect direct data or verify interpretations.

The integration of DMT with these methodologies not only amplifies its educational potential but also allows it to be adapted to different learning styles and promotes a truly interdisciplinary approach, where Earth Sciences interact naturally with other scientific, humanistic, and social disciplines.

13.5 Assessment in Data Mining Tests

Assessment in DMT requires a different approach than traditional tests, as it evaluates both the correctness of the answers and the analysis process. Effective assessment should consider:

Assessment criteria:

- **Interpretive accuracy:** Ability to correctly read graphs, maps, and data
- **Connections between information:** Ability to link data from different sources
- **Application of knowledge:** Integration of theory and new data
- **Argumentation:** Quality of explanations and support of conclusions with evidence
- **Critical thinking:** Ability to identify limitations in data and formulate alternative hypotheses

Adaptation for school levels:

Lower secondary school: Simplified datasets, guided questions, observable local phenomena.

Upper secondary school: More complex datasets, alternating between guided and open questions, simple quantitative analysis.

Upper secondary school: Complex datasets, mainly open questions, sophisticated quantitative analysis, projections of future scenarios.

Technical institutes: Professional case studies, simplified GIS software, risk assessment and interventions.

This structure allows DMT to be adapted to the different cognitive abilities and specific educational objectives of each school level.

Comparison between Traditional Assessment and DMT

Aspect	Traditional Assessment	Data Mining Test
Focus	Memorization of concepts	Analysis and interpretation
Type of knowledge	Declarative	Procedural and metacognitive
Relationship with data	Pre-processed and simplified data	Real and complex data
Cognitive process	Mainly reproductive	Mainly productive
Structure	Isolated questions	Interconnected and progressive questions
Contextualization	Often decontextualized	Strongly contextualized
Assessed competencies	Predominantly disciplinary	Disciplinary and transversal (soft skills)

DMT does not replace but integrates traditional forms of assessment,
offering a complementary approach oriented towards real data analysis

Application examples of the Data Mining Test: Regional Case Studies as Methodological Models

To illustrate the practical application of the Data Mining Test in Earth Science teaching, we present two structured examples that demonstrate how this approach can be implemented in real teaching contexts. These examples have been developed and tested in different classes, demonstrating their effectiveness in integrating theoretical knowledge with practical skills and in promoting a systemic approach to understanding geological phenomena.

Both case studies utilize sites in the Italian Alps, chosen for their accessibility, data availability, and pedagogical clarity. However, the DMT framework they illustrate is universally applicable to analogous geological contexts worldwide.

The first DMT explores climate change through glacial retreat—a phenomenon observable in glaciated regions across all continents, from the Rockies and Cascades to the Andes, from Scandinavian glaciers to those in New Zealand, the Himalayas, and East African mountains. The second DMT examines relationships between geology, morphology, and natural hazards through comparative valley analysis—an approach equally applicable to mountain regions globally, whether analysing valleys in the Appalachians, Atlas Mountains, Pyrenees, or Southern Alps.

These examples are not rigid models, but flexible frameworks that can be adapted to different educational contexts, school levels, and local realities. Teachers can use them as inspiration, modifying them according to the characteristics of their local area and specific educational objectives. Teachers should view these as templates demonstrating how to structure progressive analytical questions, integrate multiple data types, and guide students from observation through interpretation to synthesis.

What makes these examples transferable is not their location but their methodology: they show how to transform accessible data (historical photographs, topographic maps, vegetation maps, geological maps, time-series graphs) into structured investigative experiences that develop both disciplinary knowledge and analytical skills. The structure of these DMTs reflects the methodological principles discussed in the previous sections: they start from direct observation of concrete phenomena, integrate different types of data, stimulate critical thinking through progressive questions, and culminate in the development of complex hypotheses and assessments that link the geological phenomenon to broader social and environmental implications. The questioning strategies, the integration of multiple data sources, and the progression toward increasingly complex synthesis represent a framework applicable to diverse geological phenomena and educational contexts worldwide. The specific geographical content (Rutor Glacier, Lys and Ayas valleys) can be substituted with comparable features from any region while maintaining the pedagogical structure: the analytical progression from visual comparison through quantitative analysis to predictive modelling remains effective regardless of the specific glacier, valley system, or geological context being examined.

TEACHING UNIT: DMT, the case of the Rutor glacier

This DMT integrates different types of data to develop a systemic understanding of the phenomenon of glacial retreat and its environmental and social implications.

The Rutor Glacier, with an area of 8.4 km², is the third largest in the Aosta Valley and is emblematic for its geographical location and glaciological characteristics. Located on the northern slope of the mountain of the same name and facing north-northwest, this glacier occupies an area between 3,400 and 2,500 m above sea level and is divided into two sectors by the rocky spur of the Vedettes du Rutor.

Theme: Global warming and glacier retreat

Tools required: Historical images of the glacier (1919 and 2005), topographical and geological maps, graphs of the glacier's mass balance from the hydrological year 2000-2001 to 2007.

Recommended time: 45 minutes
Type of analysis:

Comparative analysis of historical images

1. Compare the images from 1919 and 2005 and identify and describe the most significant changes in the position of the glacier front and the surrounding morphology.

2. Observe the areas recently left uncovered by the glaciers: what characteristic features indicate the retreat of the glacier? What geomorphological processes are currently active in these areas?

3. The proglacial lakes visible in the images are a recent formation. Explain the mechanism of formation of these lakes and formulate a hypothesis about their possible future evolution.

4. The peat bog uncovered by the retreat of the glacier contains plant remains dating back to between 10,000 and 5,500 years ago. What information can we glean from this data about the behaviour of the glacier during the Holocene Thermal Maximum, and what parallels can we draw with the current situation?

Images of the Rutor glacier (1919, 2005)

Interpretation of topographic and geological maps

Using the map scale, measure the distance between the Deffeyes Refuge (2,494 m) and the current front of the glacier. Calculate the travel time assuming an average speed of 300 m of elevation gain per hour.

Trace the catchment area of the Rutor glacier on the map and calculate its approximate surface area. What percentage of the total surface area of the glacier is represented by the catchment area?

Identify the main lithological units present in the glacier area on the geological map and analyse how the different resistance of the rocks influences the morphology of the surrounding territory.

Locate the morainic deposits of the Little Ice Age on the map and calculate the distance between these moraines and the current position of the front, obtaining the average annual retreat rate.

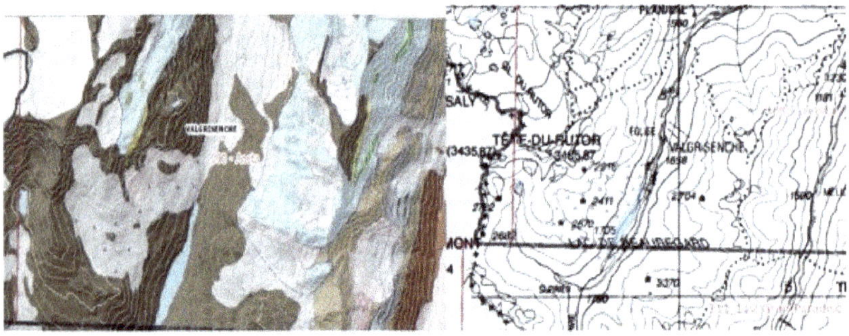

Comparison of geological map and topographic map

Analysis of time series graphs

The graph shows blue and yellow bars. Explain what these two colours represent and what relationship exists between these two components. Analyse the trend of the red line in the graph. What does this parameter represent and what does its increasingly steep downward slope mean? Identify the year with the most negative mass balance. Look up the climatic characteristics of that year in meteorological databases and explain the relationship between summer temperature anomalies, winter precipitation, and mass balance.

Looking at the temporal trend of winter accumulation and summer ablation, which of the two parameters shows greater interannual variability? Which of the two has a greater impact on the final net balance?

The graph shows the historical series of the mass balance from the hydrogeological year 2000-2001 to 2017.

The graph represents the time series of the Mass Balance from the hydrogeological year 2000-2001 to 2017 In blue: accumulation

in yellow: ablation
in green: positive balance
in red: negative balance.
Line with dots: Cumulative Balance

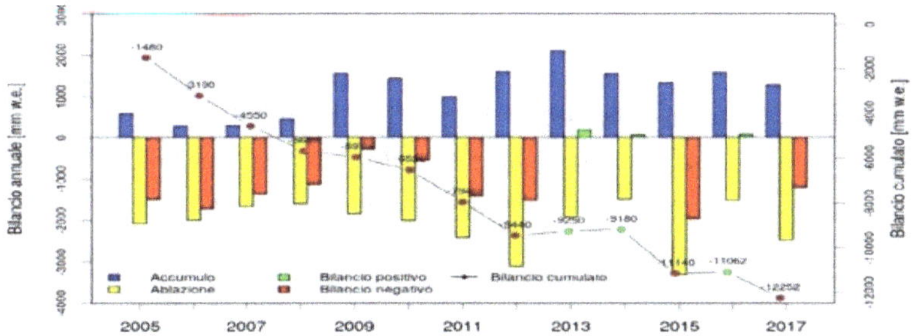

Developing hypotheses about future consequences

Considering the mass balance data and assuming that the negative trend continues at the same rate, formulate a hypothesis about the evolution of the glacier's surface area and volume over the next 30 years.
The Rutor glacier feeds important watercourses used for hydroelectric power generation. What could be the consequences of glacial retreat on the hydrological regime of these watercourses and on energy production, considering changes in both the quantity and seasonal distribution of flows?
Analyse the possible consequences of glacial retreat on the stability of slopes recently left uncovered by glaciers. What geological and morphological factors could increase the risk of landslides and debris flows in the surrounding areas?
The Rutor glacier is a destination for mountaineering and ski mountaineering. Prepare an analysis of the economic and social impacts of glacial retreat on local tourism activities and propose possible adaptation strategies.
This DMT example demonstrates how real, accessible data can be used to develop not only specific knowledge about glaciers and climate change in young people, but also analytical, synthesis, and critical evaluation skills, which are essential for addressing contemporary environmental challenges.

TEACHING UNIT: DMT, ALPINE VALLEYS- LYS AND AYAS

This Data Mining Test consists of a comparative analysis of two adjacent Alpine valleys which, despite sharing many geographical characteristics, have substantial morphological differences linked to their different geological structures. This case study offers an opportunity to explore the relationships between geology, geomorphology, vegetation cover, and the distribution of natural hazards.

The Lys and Ayas valleys, located on the orographic left side of the Aosta Valley, are two perfect examples for this type of analysis. Both are transverse valleys of the Alpine chain running north-south, located at the same latitude, with the same exposure and culminating in the north in the Monte Rosa group, whose peaks exceed 4000 meters. Despite these similarities, the two valleys have significantly different morphological characteristics that a careful observer can already see from panoramic images.

Topic: Relationship between geology, morphology, and natural hazards

Tools required: Panoramic images of the valleys, forest cover map, landslide map, geological map

Recommended time: 45 minutes

Type of analysis:

Comparative morphological analysis

Looking at the images of the two valleys, what are the main differences you notice in the morphology of the slopes? Describe the characteristics that make the profile of each valley distinctive.
The Lys valley has steeper and more rugged slopes, while the Ayas valley has gentler, more rounded slopes. What geomorphological processes could be responsible for these differences?
Analyse the width of the valley floor in the two images. Which of the two valleys has a wider valley floor and what factors could explain this difference?
Identify evidence of active geomorphological processes such as landslides, debris cones, or alluvial deposits in the images. In which of the two valleys do these phenomena appear more widespread and why?

Interpretation of vegetation distribution

Analyse the forest cover map and calculate and compare the percentage of forested area in the two valleys. What factor could explain the difference observed?
Observe the distribution of forests in relation to slope exposure. Is there an asymmetry in forest cover between east-facing and west-facing slopes? What ecological factors could explain this pattern?
Draw a line on the map representing the upper limit of the forest in the two valleys. At what average altitude is this limit located and what differences do you notice between the two valleys?

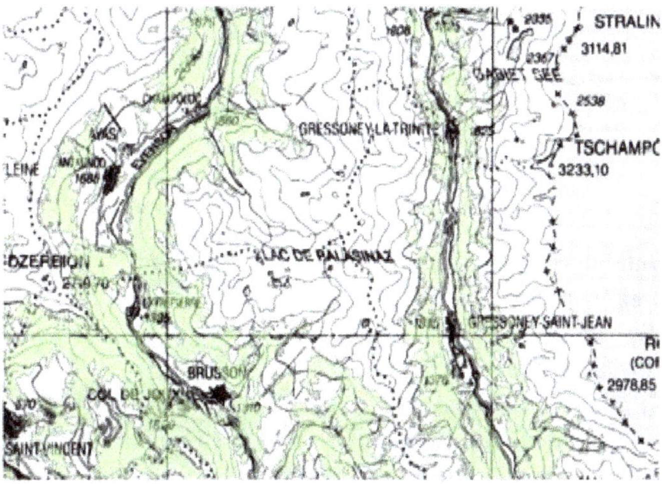

Correlate the distribution of forest cover with the morphology of the area. In which morphological contexts (steep slopes, flat areas, watersheds) does the vegetation appear sparser or absent? Formulate a hypothesis about the limiting factors.

Analysis of the distribution of landslide phenomena

On the landslide map, identify the types of phenomena prevalent in the Lys Valley and the Ayas Valley. Is there a significant difference in the distribution and type of landslide phenomena between the two valleys?
Calculate the density of landslide phenomena (number of phenomena per km²) for each valley. Which of the two has greater slope instability?
Analyse the altitudinal distribution of landslide phenomena. Is there an altitude range particularly affected by these phenomena? What factors could explain this distribution?
Mentally superimpose the landslide map on the forest cover map. What relationship can you identify between the presence of forest and landslide phenomena? Does the forest seem to play an effective protective role in both valleys?

Correlation between geology and land characteristics

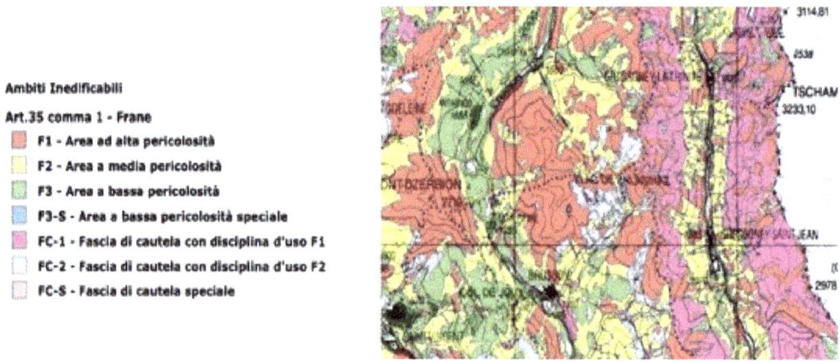

Non-Buildable Areas Art.35 paragraph 1 - Landslides
F1 - High hazard area
F2 - Medium hazard area
F3 - Low hazard area
F3-S - Special low hazard area
FC-1 - Precautionary buffer zone with F1 use regulations
FC-2 - Precautionary buffer zone with F2 use regulations
FC-S - Special precautionary buffer zone

Analyse the geological map and identify the main lithological units present in the two valleys. Which lithology is prevalent in the Lys Valley and which in the Ayas Valley?
The Lys Valley is characterized by the prevalence of crystalline rocks such as gneiss and mica schist, while prasinites and serpentinite are more widespread in the Ayas Valley. Research the mechanical properties of these rocks and explain how these characteristics can influence the morphology of the slope.
Identify the main tectonic features (faults, overthrusts) present in the area on the geological map. How do these structural elements influence the distribution of landslides?
By integrating all the data analysed (morphology, forest cover, instability, geology), construct a conceptual model that explains the differences observed between the two valleys, highlighting the causal relationships between the various factors.

The educational value of this DMT lies in understanding interconnections: students are guided to understand how basic geology influences numerous aspects of the territorial system, from landforms to vegetation distribution to the location of instability phenomena. This integrated approach overcomes the fragmentation typical of traditional Earth Science teaching, where different aspects (geology, geomorphology, hydrogeology, natural hazards) are often treated separately.
In addition, this type of DMT develops spatial analysis and integrated reading skills for different types of thematic maps, which are essential for comprehending the territory and its risks. Students learn to 'read' the landscape not only in its aesthetic dimension, but as the result of complex interactions between natural and anthropogenic components.

Iceland - The Laugahraun obsidian field

Further reading

• Bao, F., He, X., & Zhao, F. (2012). Applying Data Mining to the Geosciences Data. Physics Procedia, 33, 1401-1405. https://doi.org/10.1016/j.phpro.2012.05.229

• Occhipinti, S. (2022). The Data Mining Test: An Unexpectedly Effective Tool to Promote Soft and Hard Skills in Earth Sciences. Earth & Environmental Science Research & Reviews, 5(2), 34-43.

• Jain, P. (2021). Data Mining Tasks – Overview. Include Help. https://www.includehelp.com/basics/data-mining-tasks-overview

• Fu, Y. (1997). Data Mining: Tasks, Techniques and Applications. IEEE Potentials, 16(4), 18-20. https://doi.org/10.1109/45.624339

• Baker, R. S. (2010). Data Mining for Education. International Encyclopaedia of Education, 7(3), 112-118.

CHAPTER 14.
FORENSIC GEOLOGY: THE DETECTIVE GEOLOGIST

Forensic geology represents a fascinating intersection between Earth science and investigative inquiry. Its appeal lies in its ability to transform seemingly insignificant elements—a grain of sand, a fragment of rock, a trace of mineral—into crucial clues for solving criminal, environmental, or historical mysteries. Interest in this discipline has grown exponentially in recent years, fuelled by popular television series such as “CSI,” “Bones,” and “Elementary,” which have brought the investigative potential of the geological sciences into the homes of millions of viewers.

What makes forensic geology exceptionally intriguing is its multidisciplinary nature, which requires diverse skills ranging from mineralogy to geochemistry, sedimentology to geophysics. Each case is a unique puzzle that stimulates critical thinking and the ability to integrate information from different fields. For students, this discipline offers a valuable opportunity to see the practical and concrete application of geological concepts that are often perceived as abstract or distant from everyday life.

The versatility of forensic geology also makes it an exceptional teaching tool: it allows case studies to be built based on real events, but also fictional scenarios to be created specifically to develop analytical skills. This possibility opens the door to engaging exercises where it is possible to take on the role of 'detectives of the Earth', applying scientific knowledge in stimulating and meaningful contexts.

14.1 Definition and skills

Forensic geology is the discipline that applies the principles and methodologies of geological sciences to legal and forensic investigations. It is defined as the use of evidence based on minerals, soils, sediments, and other geological materials to assist in criminal, civil, and environmental investigations. This specialized branch integrates geology with the principles of forensic science, allowing geologists to contribute to the resolution of legal issues through the scientific analysis of terrestrial materials.

Forensic geologists bring skills from various areas of geoscience to the table. Petrography and mineralogy enable them to identify and characterize rocks, minerals, and synthetic materials, while sedimentology enables them to analyse soil particles, sand, and dust to determine their origin and history. Their knowledge of geochemistry allows them to determine the chemical composition of geological materials, including isotope analysis. They use non-invasive geophysical techniques for subsurface exploration, while their expertise in environmental geology allows them to assess contamination and its dispersion pathways. Structural geology provides them with the tools to interpret ground deformation and instability, while hydrogeology supports them to understand groundwater movements and contaminant transport.

Like a detective, the forensic geologists approach each case with a methodical and systematic approach. Thay collect fragmentary clues from often incomplete "scenes," paying attention to microscopic details that might escape the untrained eye. They can reconstruct past events based on present traces, interpreting the information preserved in geological materials. They can distinguish between natural features and alterations due to human intervention. They formulate testable hypotheses and subject them to rigorous analytical testing, progressively eliminating alternative explanations. A key skill is linking seemingly unrelated evidence into a coherent picture, establishing causal relationships between different geological elements. Finally, they must be able to communicate complex findings in a way that is understandable to non-specialists, translating technical data into narratives accessible to judges, lawyers, and juries.

The main motivation behind the use of forensic geology lies in the ability of geological materials to preserve traces of the events and environments with which they have come into contact, acting as silent witnesses to natural processes or criminal activities.

For professionals interested in this field, there are now specialized training programs such as master's degrees and postgraduate courses that integrate geological, criminological, and legal skills. Employment prospects range from forensic laboratories of the police to consulting for courts and law firms, from environmental protection agencies to specialized research institutes, to museums for the authentication of archaeological and paleontological finds. The skills required include analytical precision, communication skills, knowledge of legal procedures, and an interdisciplinary approach that integrates geology, chemistry, biology, and criminology.

14.2 Emerging technologies in forensic geology

Forensic geology is evolving rapidly thanks to the integration of new technologies that expand its investigative potential:

Advanced analysis systems

Artificial intelligence and machine learning enable the automated recognition of mineralogical patterns and rapid comparison of thousands of samples. Specialized algorithms can automatically classify soil types or mineral particles with accuracy comparable to that of human experts, but in much less time.

X-ray microtomography produces non-destructive three-dimensional visualization of samples, making it possible to analyse the internal structure of sediments and materials without compromising their integrity. The resolution of modern equipment makes it possible to visualize microscopic details, revealing structures that would otherwise be invisible.

Field instrumentation

Remote sensors and drones allow detailed mapping of hard-to-reach areas and detection of anomalies in the ground. They are particularly effective in identifying clandestine excavation sites or illegal landfills. Modern drone platforms integrate multispectral sensors that detect alterations in vegetation or soil invisible to the human eye.

Portable analysers allow detailed investigations directly in the field. Tools such as X-ray fluorescence analysers and Raman spectrometers can provide near-instantaneous compositional data without the need to transport samples to a laboratory.

Integrated approaches

Environmental DNA (eDNA) extracted from soil, sediment, or water samples represents an innovative bridge between geology and forensic biology. This technique the identification of species presents in the environment and the linking of suspects to crime scenes through biological traces.

Geographic information systems (GIS) and Big Data enable complex spatial correlations and the creation of increasingly extensive geochemical reference databases, allowing data to be visualized and analysed to reveal relationships and patterns that may not be apparent with traditional methods.

Artificial intelligence: a revolution in progress

Modern deep learning algorithms analyse microscopic images of sediments and minerals, creating detailed "geological signatures" of different environments. This facilitates quick comparisons between forensic samples and reference databases, reducing the time needed to determine the origin of a sample from days to minutes.

Machine learning algorithms excel at identifying subtle patterns in data that may escape the human eye, while predictive models allow us to simulate how geological materials would behave under different conditions, enabling us to reconstruct the timeline of criminal events.

Despite its potential, the application of AI faces significant challenges: the need for large reference databases, the interpretability of results in a legal context, and issues surrounding the admissibility in court of evidence based on complex algorithms.

14.3 Emblematic cases of forensic geology

CASE 1: THE LINDOW MOOR MURDER

Context: In 1983, a partially preserved body was discovered in the Lindow Moor (England), initially believed to be a recent victim.

Geological investigation: Stratigraphic analysis of the peat surrounding the body → Radiocarbon dating of associated tissues and plant materials → Study of pollen present in the peat layers → Stable isotope analysis to determine the victim's diet

Results: Analysis revealed that the body, later named "Lindow Man," dated back approximately 2,000 years (Iron Age). The geological study established that this was not a recent murder but an important archaeological find, probably the victim of a ritual sacrifice. Analysis of pollen and peat stratigraphy made it possible to reconstruct the environment and climate at the time of death.

Impact: This case demonstrates how forensic geology can determine not only the location but also the time of an event, distinguishing between recent crimes and archaeological contexts.

CASE 2: THE PORTO MARGHERA DISASTER

Context: In the 1990s, an environmental investigation revealed serious contamination in the sediments of the Venice Lagoon, near the industrial hub of Porto Marghera.

Porto Marghera

Geological investigation: Systematic sampling of lagoon sediments along predefined transects → Stratigraphic analysis of core samples to determine the chronology of contamination → Geochemical characterization of contaminants (heavy metals, organic compounds) → Isotopic analysis to identify specific "signatures" of pollutants → Hydrodynamic modelling to reconstruct the dispersion pathways of pollutants

Results: Geological analysis determined that contamination began in the 1950s, coinciding with industrial expansion. Researchers were able to distinguish between different sources of pollution thanks to specific isotopic "signatures," demonstrating the causal link between specific industrial activities and environmental damage. In addition, it was possible to accurately quantify the spatial and temporal extent of the pollution, providing crucial data for subsequent legal action and remediation.

Impact: The geological evidence formed the basis for legal proceedings that led to the conviction of the companies responsible and the launch of remediation programs. This case illustrates how forensic geology can be decisive in environmental litigation.

CASE 3: THE MYSTERY OF THE RIACE BRONZES

Context: The Riace Bronzes, sculptural masterpieces from ancient Greece found in the sea off Riace in 1972, have raised numerous questions about their origin and the dynamics of the shipwreck that brought them to the Calabrian seabed.

Geological investigation: Analysis of marine sediments encrusted on the statues → Mineralogical and geochemical study of metal inclusions in the statues → Comparison with the geological characteristics of potential production sites in Greece → Modelling of sea currents to reconstruct the dynamics of the shipwreck

Results: The analysis provided clues about the likely origin of the raw materials used to cast the statues, linking them to specific mining regions in ancient Greece. The study of the sediments contributed to dating the shipwreck and reconstructing the conditions in which the statues were found on the seabed.

Impact: This case demonstrates the application of forensic geology in the field of archaeology and cultural heritage, helping to solve historical mysteries and combat illicit trafficking in cultural property.

CASE 4: THE SOMERTON BEACH MURDER

Context: In 1948, an unidentified man was found dead on Somerton Beach in Australia. The case, which remains unsolved, is known as 'the mystery of the Somerton man'.

Geological investigation: Analysis of sand grains found on the victim's clothing → Mineralogical study of dust found in pockets and shoes → Comparison with reference samples from different locations in Australia → Identification of microfossils present in sediments

Results: Geological analysis revealed that some minerals and sediments found on the man's clothing did not match those found on Somerton Beach, suggesting that the victim had visited other areas prior to his death. Specific mineralogical associations provided clues about the places he had visited in the hours before his death.

Impact: Although the case remains officially unsolved, the geological investigation provided crucial information about the victim's movements, demonstrating how even in 'cold' cases, geology can offer new investigative perspectives.

CASE 5: FOSSIL TRAFFICKING FROM MONGOLIA

Context: In 2012, US authorities seized an almost complete skeleton of Tarbosaurus bataar, a dinosaur like T-rex, which was about to be auctioned for over a million dollars.

Geological investigation: Analysis of the rock matrix still adhering to the fossils → mineralogical and geochemical characterization of the sediment → study of associated microfossils → comparison with reference samples from the Nemegt Formation in Mongolia

Results: Geological analysis unequivocally demonstrated that the fossil originated from the Nemegt Formation in Mongolia, from which the export of fossils is illegal. The unique characteristics of the sedimentary matrix, including its mineralogical composition and specific association of microfossils, provided an irrefutable "geological fingerprint."

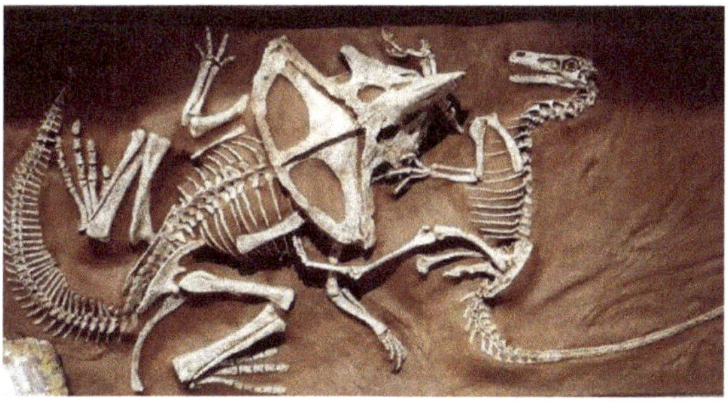

Fossil record of Mongolia

Impact: Based on the geological evidence, the court ordered the fossil to be returned to Mongolia and sentenced the trafficker to three months in prison and confiscation of assets. This case set an important precedent for the use of forensic geological techniques in the fight against illicit trafficking in paleontological assets.

CASE 6: THE CASE OF THE FAKE PAINTER

Context: An internationally renowned art gallery discovered that some works attributed to a famous 19th-century landscape painter had suspicious characteristics. An investigation was requested to determine the authenticity of the paintings.

Geological investigation: Microscopic sampling of mineral pigments → spectroscopic analysis of the components → dating of the organic elements present → comparison with the known geological composition of pigments of the period → reconstruction of the provenance of the minerals used

Results: The analysis revealed the use of pigments containing minerals whose commercial extraction only began in the second half of the 20th century, making them incompatible with the declared age of the works. In addition, the isotopic proportions of certain trace elements differed significantly from those found in authentic works by the artist.

Impact: The geological evidence made it possible to identify a sophisticated forgery system, leading to the recovery of several falsely attributed works with a market value of several million euros. This case highlighted the value of forensic geology in the field of cultural property authentication.

CASE 7: THE MYSTERY OF THE POISONED LAKE

Context: In a remote mountain region, an alpine lake suddenly showed signs of severe pollution with fish deaths and chemical alteration of the water. Local authorities suspected illegal industrial waste disposal, but in the absence of witnesses, it was difficult to identify those responsible.

Geological investigation: Analysis of lake bottom sediments → chemical and mineralogical characterization of pollutants → study of recent stratifications → hydrogeological tracing of water flows in the basin → modelling of contaminant dispersion → comparison with the chemical "signatures" of potential industrial sources in the area

Results: Contrary to initial assumptions, geological analysis revealed that the contamination did not come from direct discharges into the lake, but from an old abandoned mine located upstream, where recent excavation work for a tourism project had exposed Sulphur minerals. The oxidation of these minerals had generated acidic water which, infiltrating the subsoil, had reached the lake through underwater springs, releasing heavy metals and altering the pH of the ecosystem.

Impact: The case demonstrated how forensic geology can not only identify the cause of pollution, but also correct erroneous investigative hypotheses, avoiding the prosecution of innocent parties and correctly directing mitigation measures. The company responsible for the excavation was forced to remediate the mining area and implement systems for the collection and treatment of contaminated water.

CASE 8: THE VIETNAM WAR - A FORENSIC GEOLOGICAL INVESTIGATION IN THE MILITARY FIELD

Context: During the Vietnam War (1955-1975), one of the Vietcong's main tactical advantages was a complex system of underground tunnels that allowed them to hide and emerge suddenly for surprise attacks.

Geological investigation: Systematic study of the geological characteristics of the "iron triangle" northwest of Saigon → Analysis of the soil structure and identification of sedimentary formations favourable to excavation → Mapping of areas of potential extension of underground networks based on geotechnical data

Results: The investigations led to Operation Cedar Falls in 1967, with the removal of surface vegetation to identify camouflaged entrances and the use of heavy equipment in geologically compatible areas.

Impact: Despite the application of advanced geological expertise, approximately 60% of the tunnel system remained intact, demonstrating the limitations of geological intelligence when compared to empirical local knowledge of the area. This case illustrates how soil analysis can be used to "read" the terrain in search of hidden human interventions, a skill now applied in numerous civil investigative contexts.

Operation Cedar Falls
part of the Vietnam War

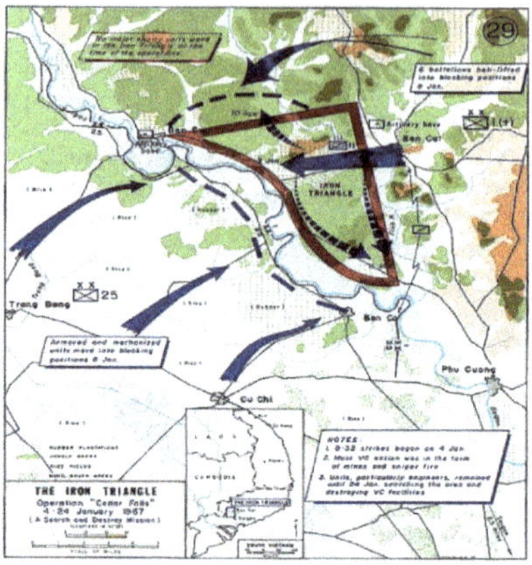

Map of the territory involved in
Operation Cedar Falls

January 8-26, 1967

14.4 Emerging applications and prospects of forensic geology

Forensic geology is rapidly expanding its application horizons, embracing increasingly diversified fields ranging from national security to cultural heritage protection. In the field of terrorism and security, the analysis of geological materials presents in explosive devices facilitate the identification of raw material provenance and the reconstruction of supply networks. Similarly, the fight against food fraud increasingly relies on trace element analysis of soil to verify the declared geographical origin of products, a valuable expertise above all for PDO and PGI products.

Technological evolution is revolutionizing investigative methodologies. The use of stable isotopes as "geographical fingerprints" now makes possible precise tracing of the movements of people, animals, and objects across continents, providing decisive evidence in cases of drug trafficking and poaching. Artificial intelligence algorithms, trained on thousands of geological samples, can now identify in minutes patterns that would require days of manual analysis, drastically accelerating investigation timelines.

A growing challenge concerns the legal admissibility of evidence based on complex algorithms, which requires the development of standardized protocols and the training of experts capable of explaining technical results to judges and juries. The miniaturization of analytical instrumentation promises to bring portable laboratories directly to crime scenes, while the creation of interconnected global geochemical databases will facilitate increasingly precise and rapid comparisons.

The future of the discipline is oriented toward an increasingly interdisciplinary approach, where forensic geology integrates with molecular biology, archaeology, and cybersecurity. Increasing digitalization will require new professional skills that combine traditional geological knowledge with advanced computer skills, opening unprecedented career prospects for new generations of geologists.

Forensic geology demonstrates that the most solid evidence is literally "carved in stone," and its investigative approach represents a powerful educational tool for developing observation and critical thinking skills essential for facing future challenges.

Columnar rhyolites in Yellowstone Park

Further reading

Di Maggio, R.M., & Barone, P.M. (2018). *Geologia Forense. Flaccovio Editore*. https://www.flaccovio.com/geologia-forense-p-1029.html

DiMaggio, R.M., & Donnelly, L.J. (2012). "*The Application of Forensic Geology to Homicide Investigations.*" Australian Journal of Forensic Sciences, 44(2), 205-221. https://doi.org/10.1080/00450618.2011.636798

Donnelly, L. (2011). "*The Role of Geoforensics in Policing and Law Enforcement.*" Emergency Services Times, 12(1), 49-52.

Murray, R.C. (2011). *Evidence from the Earth: Forensic Geology and Criminal Investigation*. Mountain Press Publishing Company. https://mountain-press.com/products/evidence-from-the-earth

CHAPTER 15.
GEOSCIENCES AND COMPETENCIES: AN EFFECTIVE RELATIONSHIP

The traditional definition of competence comes from educational research and is defined as "the achievement of a performance in a given context that involves the use of attitudes and motivations, knowledge, skills, and abilities and is aimed at achieving a goal." More precisely, competence is "what, in a given context, can be done (ability) on the basis of knowledge to achieve the intended goal and produce knowledge." It involves choosing, using, and mastering appropriate knowledge, skills, and abilities in each context to set and/or solve a specific problem.

The competency-based approach to teaching geosciences represents a paradigm shift from the traditional transmission of disciplinary content. Competencies are not simply technical skills or specific knowledge, but complex cognitive configurations that integrate knowledge, procedural skills, and personal attitudes.

In geosciences, this approach is notably significant because the very nature of geological phenomena—characterized by multiple interdependencies, different temporal and spatial scales, and complex causal chains—provides an ideal terrain for the development of transversal competencies. The Earth thus becomes not only an object of study, but also a powerful cognitive mediator that stimulates the acquisition of intellectual tools applicable well beyond the geological context.

Teaching experience has shown that geosciences are particularly effective in promoting transversal and citizenship skills, developing in particular the ideas of systems and complexity. This is precisely because geosciences offer an extraordinarily fertile ground for the study of complex and interconnected natural phenomena. The investigative approach typical of geology enables to acquire scientific investigation methods, systemic analysis skills, and critical thinking that can be applied far beyond the specific disciplinary context.

The distinctive features of the geological approach

The geological approach develops skills through simultaneous operation across:

Multiple time scales: From instantaneous processes such as earthquakes to phenomena that develop over millions of years such as orogenesis, geosciences accustom students to thinking about time scales that transcend ordinary human experience.

Different spatial scales: From the microscopic observation of the crystal structure of minerals to the analysis of global tectonic dynamics, developing extraordinary cognitive flexibility in the transition between levels of observation.

Complex systems: Geological phenomena are the result of complex interactions between the geosphere, atmosphere, hydrosphere, and biosphere, requiring a systemic approach that recognises interconnections and feedback mechanisms.

Investigative method: The impossibility of directly observing many geological processes naturally develops inferential thinking, the ability to formulate hypotheses, and interpret indirect evidence.

Geosciences develop skills that prove useful in many contexts and disciplines, precisely because of the inherently complex and cross-cutting nature of geological phenomena. This chapter presents ten key skills, each accompanied by concrete case studies illustrating how geological experience can be transferred to other areas of knowledge and human action.

For each skill, two case studies are presented that demonstrate how geological phenomena offer unique learning opportunities, followed by an analysis of how these experiences can be translated into transferable skills. We conclude with practical guidelines for the progressive development of these skills across different school levels and tools for their assessment.

The following sections present ten basic competencies developed through geoscience education, each illustrated with two concrete case studies. These examples draw from diverse geographical and temporal contexts—from local Italian geological features to global phenomena, from contemporary environmental challenges to deep geological time. While some case studies reference specific regional contexts (such as the Po Valley system or Fumane Cave in the Italian Alps), the competencies they develop are universally applicable. Teachers worldwide can complement these examples with analogous cases from their own regional geology: every continent offers comparable opportunities to develop systemic thinking through river systems, understand cause-effect relationships through volcanic hazards, explore multiscale thinking through local geological features, and foster environmental awareness through regional geological services. The pedagogical value lies not in the specific geographical location but in demonstrating how geological phenomena—regardless of where they occur—provide exceptional opportunities for developing transferable cognitive skills essential for navigating complexity in any field of human endeavour.

15.1 Systemic analysis and recognition of interconnections

The ability to identify components, processes, and interconnections is fundamental in addressing complex problems. Students learn to recognise flows and links between different elements, an essential skill in fields ranging from economics to ecology.

CASE 1: The Po River and the Po Valley. The Po River system is an ideal laboratory for discovering the interconnections between geological, hydrological, and anthropogenic elements. The lithological differences between the upper and lower plains create contrasts in permeability that determine the emergence of the line of resurgences. This strip of land with unique hydrological characteristics has profoundly influenced the development of agricultural practices, such as the "*marcite*" developed by Cistercian monks in the 13th century. By exploiting the waters of the resurgence, which maintain a constant temperature of 9-12°C, they created irrigated meadows even in winter. The evolution of the Delta, with the silting up of ancient ports such as Spina (now 12 km from the sea), demonstrates the sedimentary dynamics that continue to modify the landscape.

Similar systemic interconnections characterize river systems worldwide: the Mississippi Delta's interaction with Gulf Coast ecosystems and human settlements, the Ganges-Brahmaputra system's influence on South Asian agriculture and culture, the Nile's historical role in Egyptian civilization, or the Amazon's planetary-scale biogeochemical cycling. Each major river system offers comparable opportunities to understand how geological substrates, hydrological processes, and human activities create complex, interconnected territorial systems.

CASE 2: Critical zone. The concept of the "critical zone"—the dynamic interface where the atmosphere, biosphere, hydrosphere, and geosphere interact—offers an ideal paradigm for developing a systemic vision. The most important processes for life take place in this thin planetary film. Biogeochemical cycles connect the organic and inorganic worlds, water flows transport nutrients through the soil and subsoil, and biological activity alters the chemistry of rocks, accelerating erosion and pedogenesis. These interactions determine soil fertility, water quality, and territorial resilience.

Systemic analysis skills allow us to trace complex causal chains and understand the feedback mechanisms between different components of the system. They enable us to recognise that the boundaries between traditional terrestrial "spheres" are artificial constructs that often hinder our perception of real phenomena, allowing us instead to see the Earth system as a continuum of interconnected processes. This integrated view helps us understand how disturbances in one component can propagate throughout the entire system, creating often unpredictable cascading effects.

15.2. Understanding cause-effect relationships

Geosciences develop the ability to recognise complex causal chains in natural systems, which is essential for problems such as climate change or hydrogeological instability.

CASE 1: Yucatan cenotes and the Maya civilization Cenotes demonstrate how geological interactions can determine the development of civilizations. The Yucatan Peninsula, composed of porous limestone, has been shaped by karst processes, creating a system of underground caves. The collapse of the ceilings forms cenotes—natural wells that provide access to the aquifer. In a region with no surface rivers, these wells were the only reliable source of fresh water, determining the geographical distribution of Mayan cities. The Mayans integrated these geological features into their worldview, considering them sacred portals to the underworld where they performed rituals and sacrifices.

CASE 2: Volcanic risk in the Phlegraean Fields. The Phlegraean Fields are a complex example of the interaction between volcanic risk and urban development. This active caldera, measuring 12-15 km, is characterized by bradyseism—vertical movements caused by variations in magmatic pressure. The Serapeum of Pozzuoli shows signs of submersion and re-emergence: the lithodomian holes on the columns bear witness to periods of marine submersion. Significant bradyseismic crises occurred in 1969-72 and 1982-84, the latter with a 1.8-meter uplift that led to the evacuation of 40,000 people. Today, around 500,000 people live in the caldera area.

Recognising cause-and-effect relationships in natural systems facilitates us to develop deductive reasoning, which enables us to connect different observations into a coherent picture. It allows us to recognise non-linear cause-effect relationships and analyse the complex relationships between geological phenomena and sociocultural development. This ability to integrate knowledge from different disciplines prepares us to face contemporary challenges such as sustainable water resource management, land-use planning in geologically sensitive environments, natural risk mitigation.

Similar relationships between volcanic hazards and urban development exist globally: Long Valley Caldera near Los Angeles, Taupo Volcanic Zone in New Zealand housing multiple cities, Yellowstone Caldera underlying a major tourist destination, or Rabaul Caldera in Papua New Guinea with its rebuilt city. Each context demonstrates how geological knowledge must inform land-use decisions in volcanically active regions worldwide.

15.3. Multi-scale thinking

Geosciences require a constant shift from the microscopic to the planetary, developing cognitive flexibility that is valuable in many scientific fields where different levels of analysis need to be integrated.

CASE 1: Diamonds - messengers from deep within the Earth. Diamonds allow us to connect processes on vastly different scales. At the microscopic level, their crystal structure reveals the atomic configuration of carbon under high pressure, with tetrahedral covalent bonds that give them exceptional hardness. Trapped mineral inclusions act as 'time capsules', preserving materials from inaccessible depths. Recent studies have identified stable inclusions only in the lower mantle (over 660 km), demonstrating that some diamonds form at much greater depths. On a regional scale, they provide information on formation in kimberlites, while globally they offer clues about the Earth's internal structure and the geological carbon cycle.

CASE 2: Hydrogeological cycle and water resources. The hydrogeological cycle connects different environmental compartments, influencing the Earth system. At the microscopic level, the properties of water—molecular polarity, hydrogen bonds, anomalous freezing behaviour—determine its role as a universal solvent and transport agent. These properties influence rock alteration, nutrient transport, and water-organism interactions. On a local scale, the cycle manifests itself differently depending on the geological context: in karst regions, it creates underground systems with caves and springs, while in arid regions, it generates ephemeral flows that shape the landscape. Globally, it is a major mechanism of energy transfer, influencing the climate and planetary ecosystems.

Multiscale thinking allows us to visualize connections between processes operating on vastly different scales, from atomic structure to global dynamics. It helps us to deduce large-scale processes from the observation of microscopic characteristics and to integrate analyses conducted at different scales into a coherent interpretative framework. This "mental zoom capability" develops an integrated view of natural phenomena, understanding how microscopic properties can determine macroscopic behaviour and how global processes can manifest themselves in local characteristics.

15.4 Cause-effect analysis on different time scales

Geological phenomena operate on extremely variable time scales, from seconds to millions of years. This expertise is invaluable in addressing contemporary issues such as climate change.

CASE 1: Meteorites and the formation of the Earth's crust. During the Hadean (over 3.8 billion years ago), meteorite bombardment had effects on different time scales. The immediate impacts caused partial melting of surface rocks and the formation of giant craters. In the medium term, they altered the primordial landscape, contributing to the formation of ocean basins. The long-term effects were even more significant: they promoted crustal differentiation and the development of tectonic mechanisms that shaped the planet for billions of

years. They also brought essential elements for life such as water and organic compounds.

CASE 2: Snowball Earth. About 770 million years ago, Earth went through the global glaciation known as "Snowball Earth" with average temperatures between -20 and -50°C. In the short term, the formation of sea ice increased the Earth's albedo, intensifying cooling through positive feedback. In the medium term, the frozen planet blocked biogeochemical cycles with consequences for primitive ecosystems. In the long term, volcanoes continued to erupt, releasing CO2 and methane. Without plants or exposed oceans to absorb these greenhouse gases, their concentration gradually increased until they triggered a greenhouse effect which, after millions of years, allowed the planet to emerge from extreme glaciation.

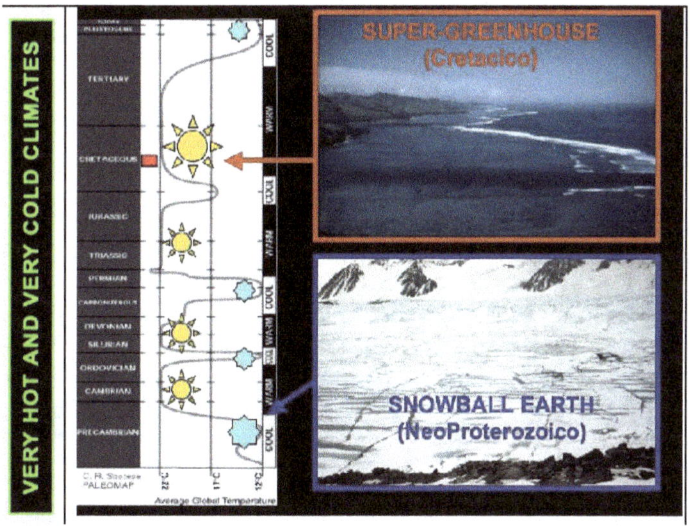

Being able to analyse processes on different time scales makes possible to analyse chains of events on geological time scales, understand how catastrophic events can trigger long-term evolutionary processes, and recognise the remote origins of current geological phenomena. This ability to consider simultaneously processes operating on different timescales develops a conception of time that transcends ordinary human experience. Such an extended view of time is essential for addressing contemporary challenges that require long-term consideration, such as the sustainable management of non-renewable resources and the assessment of climate change in the context of natural variability.

15.5. Understanding the cyclical nature of natural processes

From biogeochemical cycles to the rock cycle, geosciences teach us to recognise cyclical patterns characteristic of natural systems. This vision is essential for understanding sustainability and developing circular economic models.

CASE 1: Quaternary glacial cycles. The glacial cycles of the last 2.5 million years are an extraordinary example of natural cyclicity. The alternation of glacial and interglacial periods has shaped the European and North American landscape, creating U-shaped valleys, glacial lakes, and moraine deposits. This cyclicity follows variations in Earth's orbital parameters (Milankovitch cycles) that modify the distribution of insolation. During cooling, the expansion of ice caps increases albedo, further intensifying cooling. Glacial advance shapes the landscape through erosion, transport, and deposition. During interglacial periods, retreat exposes landscapes that undergo degradation until the next advance.

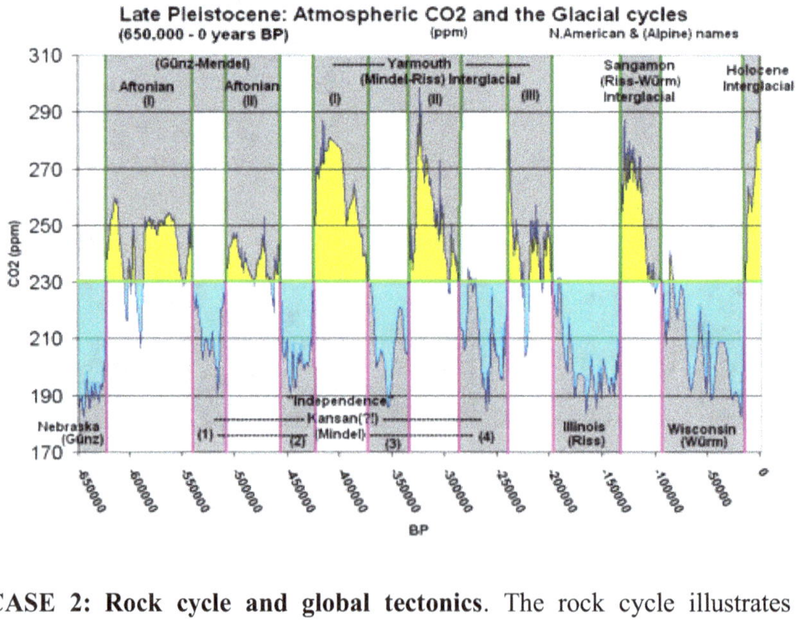

CASE 2: Rock cycle and global tectonics. The rock cycle illustrates the continuous transformation of terrestrial materials through interconnected processes linked to plate tectonics. Magma in the mantle rises and crystallizes to form igneous rocks. Exposed to atmospheric agents, they undergo alteration and disintegration. Erosive products are transported and deposited as sediments, which through burial and diagenesis become sedimentary rocks. Further burial produces metamorphism, transforming them into metamorphic rocks. Higher

temperatures and pressures can cause partial melting, generating new magma. Ophiolites—fragments of oceanic crust incorporated into mountain ranges—are a tangible example of this cycle.

Recognising cyclicality in natural processes allows us to identify recurring patterns in geological records and understand the feedback mechanisms that drive these natural cycles. It helps us to interpret the traces left by cyclic processes in the landscape and develop a deep awareness of the dynamic nature of terrestrial materials: even rocks, seemingly immutable on the human timescale, are subject to continuous transformation on the geological timescale. This cyclical view provides a powerful conceptual model for interpreting geological history and offers the conceptual basis for the development of circular economic pattern in which resources are continuously recycled rather than consumed linearly.

15.6 Critical thinking and problem solving

The investigative approach of geosciences, where clues are interpreted and hypotheses are formulated, develops rigorous critical thinking applicable to any context requiring complex data analysis.

CASE 1: Oviraptor - a scientific review. The story of Oviraptor demonstrates how scientific interpretations evolve. In 1923, Roy Chapman Andrews discovered the remains of a small theropod dinosaur crouching over a nest of eggs near a Protoceratops. The immediate interpretation was that it was stealing the eggs to eat them, leading to the name "Oviraptor philoceratops" (egg thief who loves ceratopsians). For seventy years, it was depicted as an egg predator in textbooks. It was not until the 1990s that new excavations revealed the truth: Oviraptor was protecting its eggs by displaying parental care behaviours like modern birds.

CASE 2: Geological risks and land-use planning. Assessing geological risks requires sophisticated problem solving that integrates geological data, probabilistic analysis, and socioeconomic considerations. The process includes identifying areas at risk through geological and geomorphological analysis, assessing the probability and intensity of events using historical data and mathematical representations, analysing the vulnerability of settlements considering population density and building types, and developing mitigation strategies through structural and non-structural interventions such as warning systems and land use regulations.

Critical thinking and problem solving enable us to understand that scientific interpretations are provisional and subject to revision in the light of new data, that biases can influence even analysis that is presumed to be objective, and that science progresses through a continuous process of verification. It also develops the ability to analyse complex problems with multiple variables, to integrate scientific data into decision-making, and to balance scientific, economic, and social considerations. This critical and methodical approach is a particularly valuable intellectual tool in today's age of information overload and the spread of misinformation.

15.7 Spatial interpretation

The ability to read representations of space (topographic and geological maps, satellite images, 3D models) is a basic cognitive skill that strengthens spatial intelligence.

CASE 1: Geological mapping. Geological mapping combines spatial and temporal representation, visualizing geological time on a two-dimensional plane. A geological map contains thousands of pieces of information organized according to precise codes: contour lines for altimetry, symbols for vegetation, roads, settlements, and water. Interpretation requires decoding symbols and colours representing physical and temporal characteristics, three-dimensional visualization of geological structures from two-dimensional representations, reconstruction of geological history by analysing the arrangement and relationships between units, and prediction of unobservable characteristics such as the continuation of layers at depth or the presence of aquifers.

CASE 2: Complex geological structures. The interpretation of complex structures such as folds, faults, and intrusions requires sophisticated spatial visualization skills. Geologists must "see" what is not directly observable, reconstructing three-dimensional geometries from partial information. This includes reconstructing fault systems from limited outcrops, interpreting subsurface data such as seismic profiles, and predicting the geometry of invisible layers. Particularly challenging is the analysis of areas where different tectonic events have overlapped, creating intricate deformation patterns that require reconstruction of the chronological sequence of the deformative events.

Spatial interpretation enables the development of skills in two-dimensional representation of three-dimensional realities, active spatial thinking, interpretation of symbols and cartographic conventions, and understanding of the temporal stratification of the territory. It supports the mental visualization and manipulation of three-dimensional objects, the interpolation of fragmentary information, and the perception of complex geometric relationships. This form of spatial intelligence is valuable especially in those professional fields that require the interpretation of complex spatial data, from architecture to urban planning, civil engineering to computer programming.

15.8. Long-term temporal vision

Few disciplines such as geosciences develop the ability to consider the long term in decision-making, a crucial skill for global challenges such as climate change.

CASE 1: Fumane Cave and human evolution. The Fumane Cave in the Lessini Mountains documents the presence of Neanderthals and early Homo sapiens over tens of thousands of years. This site, part of a European network of 13 Ice Age sites, allows evolutionary processes to be analysed on timescales different from everyday experience. The stratigraphic sequence records climate change over tens of thousands of years, faunal remains document ecosystem evolution, and stone tools testify to technological and cultural adaptations.

Stratigraphy of the Fumane cave

The presence of different types of humans remains illustrates dynamics of replacement and possible hybridization between human species, suggesting that "the differences between Neanderthals and Sapiens seem to be diminishing."

Archaeological sites worldwide offer similar opportunities to develop long-term temporal perspectives: Blombos Cave in South Africa documenting early modern human behaviour, Denisova Cave in Siberia revealing human species diversity, Paisley Caves in Oregon illuminating early American settlement, or Pinnacle Point in South Africa preserving evidence of coastal adaptation. Each site provides windows into deep human history that develop skills in long-term temporal reasoning.

CASE 2: Radioactive waste management. Radioactive waste management requires an extremely long-term vision: some isotopes remain dangerous for tens of thousands of years. Solutions involve deep geological confinement in stable formations, requiring an assessment of geological stability over 100,000 years or more. Geologists must predict possible changes over this time frame: climate variations that could lead to glaciation or sea level rise, tectonic movements that would alter stability, and changes in hydrogeological systems. A fascinating aspect is the need to develop hazard communication systems that can be understood by future generations who may not share the current language or symbols.

The long-term temporal perspective enables us to interpret changes on different time scales, understand the complex interactions between climate, environment, and culture over the long term, and recognise evolutionary patterns through fragmentary evidence. It supports us to project analyses over very long-time scales, understand the limits of prediction in complex systems, and consider intergenerational responsibility in current decisions. This extended temporal perspective is fundamental for addressing numerous contemporary challenges, from environmental sustainability to infrastructure planning, developing a perspective that helps contextualize current changes and promoting an ethic of responsibility towards future generations.

15.9 Environmental awareness and ecological responsibility

The study of geosciences develops a deep knowledge of the interactions between the geosphere, biosphere, hydrosphere, and atmosphere, fostering environmental awareness based on solid knowledge.

CASE 1: The human footprint in sediments Stratigraphic markers of human activity offer concrete understanding of anthropogenic impacts on the Earth system. These include microplastics in marine sediments found in the most remote sea beds, radioactive isotopes from nuclear tests that have left a global stratigraphic signal since the mid-20th century, alterations in biogeochemical cycles such as increased nitrogen in sediments from the use of fertilizers, and changes in atmospheric composition recorded in ice cores showing a rapid increase in greenhouse gases since the industrial revolution. These markers represent a "signature" of human activities recognizable in the geological record for millennia, to the point that many geologists propose recognising the Anthropocene as a new geological epoch.

CASE 2: Geological ecosystem services Geological ecosystem services—benefits that humans derive from geological processes—offer an important perspective for developing environmental awareness. They are divided into regulating services (groundwater filtration, carbon sequestration in sediments), provisioning services (minerals, fossil fuels, water resources), supporting services (geological substrate for ecosystems and nutrient cycles), and cultural services (aesthetic, recreational, and educational value of geological landscapes). Students understand that the value of geological resources goes beyond market price and that sustainable management is essential for human well-being. The analysis of geosites, places of geological interest that deserve protection for their scientific value and contribution to the natural and cultural heritage, is significant.

Environmental awareness and ecological responsibility enable us to recognise the impact of human activities on a geological scale, to understand the interconnections between everyday decisions and global processes, and to become aware of the recovery times of natural systems compared to the times of alteration. It facilitates us to assess the economic and non-economic value of geological processes, recognise human dependence on non-renewable resources, and develop the need to manage geological resources sustainably. This integrated view of the relationship between humanity and the Earth system is essential to developing an environmentally conscious and responsible citizenship, providing a basis for critically assessing the impacts of human activities and promoting sustainable practices.

15.10. Effective scientific communication

The need to communicate complex phenomena on scales that are difficult to visualize develops sophisticated communication skills that are essential in any professional field.

CASE 1: Earthquake risk communication. Earthquake risk communication requires the translation of complex scientific concepts into understandable information. It includes explaining hazards, vulnerability, and risk in accessible terms, communicating uncertainties in seismic forecasts without generating alarmism or underestimation, balancing scientific rigor and communicative effectiveness using metaphors that make complex phenomena understandable, and adapting the message to different audiences from local authorities to the general population to the media.

CASE 2: Dissemination of geological heritage. The enhancement of geological heritage requires skills to make complex concepts accessible. Geoparks are excellent examples of effective geological communication through trails with explanatory panels using understandable language supported by illustrations and reconstructions, educational materials suitable for different educational levels, and interactive technologies such as augmented reality to visualize geological history superimposed on the current landscape. Particularly effective is the narration of 'geological stories' that link natural phenomena to human experience, such as the story of a river that shaped landscapes and influenced settlements.

Effective scientific communication allows the development of the ability to simplify complex concepts without trivializing them, to communicate scientific uncertainty effectively, and to create understandable visualizations of data and phenomena. It makes possible technical language to be adapted to different communication contexts, metaphors and analogies to be used creatively to explain complex concepts, effective visual aids to be created, and narrative skills to be developed to make geological topics engaging. These communication skills, which balance scientific rigor and accessibility, are essential in many professional contexts and represent a fundamental skill for dialogue between science and society.

15.11 The network of skills: cognitive tools for a complex world

The skills explored illustrate how geosciences are a fertile ground for the development of cognitive skills that transcend disciplinary boundaries. These skills do not operate in isolation but form an interconnected network that enhances students' ability to tackle complex problems in any field. Together, these skills constitute real 'cognitive tools' that enrich thinking and action in any field of knowledge. When geared towards the development of these skills, geoscience education thus becomes a powerful vehicle for comprehensive education, preparing them to face the complex challenges of a rapidly changing world with robust and flexible conceptual tools.

15.12 Developing skills: from theory to teaching practice

The skills outlined above do not develop automatically through the simple transmission of content but require a targeted teaching approach that integrates disciplinary knowledge, concrete experiences, and metacognitive reflection.

The assessment of skills requires a different approach than traditional knowledge testing, using authentic tasks in real situations, process-based assessment of reasoning beyond the result, self-assessment for metacognitive awareness, and constructive feedback aimed at improvement.

COMPETENCIES AND DIDACTICS IN GEOSCIENCES

PRIMARY SCHOOL (6-11 years)

Competency	Concrete Activities	Geological Contents	Assessment
Systemic Analysis	• Ecosystem terrarium (4 weeks) • Soil-vegetation observation	• Water cycle, soil formation • Rock-plant interactions	• Recognizes 3 system elements • Describes 2 changes • Formulates 1 connection
Cause-Effect Relations	• Erosion experiments with containers • Seasonal cycle observation	• Erosive processes • Vegetation protection	• Connects rain-erosion • Predicts simple effects
Multi-scale Thinking	• Geological timeline (10 meters) • Fossil exploration	• Geological time • Evolution of life	• Compares different durations • Understands "very long ago"
Temporal Analysis	• Observation of slow/fast processes • Rock formation sequences	• Speed of geological processes • Stratification	• Distinguishes fast/slow • Orders temporal sequences

Competency	Concrete Activities	Geological Contents	Assessment
Process Cyclicity	• Seasonal observation • Simple natural cycles	• Day/night alternation • Seasonal cycles	• Recognizes repetitions • Predicts simple cycles
Critical Thinking	• Rock classification • Guided investigations	• Mineral properties • Rock characteristics	• Compares characteristics • Formulates simple hypotheses
Spatial Interpretation	• Simple territory maps • Plastic models	• Space representation • Landscape forms	• Reads simple maps • Builds models
Temporal Vision	• Simple geological stories • Past-present comparison	• Landscape changes • Traces of the past	• Recognizes changes • Distinguishes before/after
Environmental Awareness	• Local environment observation • Differentiated collection	• Natural resources • Environmental protection	• Recognizes resources • Respectful behaviors
Communication	• Posters with tactile materials • Small class exhibitions	• Rock and mineral properties • Landscape forms	• Uses simple language • Explains to companions

SECONDARY SCHOOL I GRADE (11-14 years) - Advanced

Competency	Concrete Activities	Geological Contents	Assessment
Systemic Analysis	• Hydrographic basin mapping • Hydrological cycle simulations	• Hydrographic basins • Geology-hydrology	• Identifies components • Traces connections • Predicts consequences
Cause-Effect Relations	• Local geological events analysis • Topography-microclimate study	• Landslides, floods • Relief influence on climate	• Reconstructs causal chains • Identifies multiple factors
Multi-scale Thinking	• Proportional time models • Slow process analysis	• Geological time scales • Landscape evolution	• Represents different scales • Connects micro-macro
Temporal Analysis	• Stratification study • Geological events chronology	• Stratigraphy principles • Relative dating	• Reconstructs sequences • Applies dating principles
Process Cyclicity	• Glacial cycles • Periodic phenomena analysis	• Quaternary glaciations • Sedimentary cycles	• Identifies periodicity • Predicts evolutions

Competency	Concrete Activities	Geological Contents	Assessment
Critical Thinking	• Geological risk news analysis • Diverse source comparison	• Natural risks • Data interpretation	• Distinguishes facts-opinions • Evaluates sources • Formulates questions
Spatial Interpretation	• Local landscape form observation • Territory plastic model construction	• Erosive and depositional forms • Topographic map reading	• Recognizes main forms • Builds simple models
Temporal Vision	• Territorial history reconstruction • Historical change analysis	• Local geological evolution • Anthropic modifications	• Connects past-present • Predicts evolutions
Environmental Awareness	• Anthropic impact analysis • Sustainability study	• Pollution, caves • Resource use	• Evaluates impacts • Proposes solutions
Communication	• Geological podcasts/videos • Companion guides	• Local geological phenomena • Territory history	• Adapts language • Structures logic • Involves audience

SECONDARY SCHOOL II GRADE (14-19 years)

Competency	Concrete Activities	Geological Contents	Assessment
Systemic Analysis	• GIS territory projects • Interdisciplinary analysis • Complex system modeling • Geological big data integration	• Hydrogeological catastrophe • Complex spatial analysis • Earth systems • Geological databases	• Integrates multiple data • Models scenarios • Proposes solutions • Manages complexity
Cause-Effect Relations	• Field investigations • Complex case analysis	• Yucatan cenotes • Campi Flegrei bradyseism	• Reconstructs complex chains • Integrates multiple factors
Multi-scale Thinking	• Glacial cycle research • Evolution simulations	• Climate cyclicity • Planetary processes	• Integrates spatio-temporal scales • Models systems
Temporal Analysis	• Paleoclimatic analysis • Event reconstruction	• Climate proxies • Extreme past events	• Analyzes time series • Correlates distant events
Process Cyclicity	• Biogeochemical cycle study • Periodicity analysis	• Milankovitch cycles • Sedimentary cycles	• Models feedback • Predicts evolutions

Competency	Concrete Activities	Geological Contents	Assessment
Critical Thinking	• Problem-Based Learning • Controversial debates • Scientific literature analysis • Peer-review assessment	• Climate changes • Resource management • Geological research • Scientific methodology	• Analyzes evidence • Arguments positions • Mediates between visions • Evaluates research quality
Spatial Interpretation	• Complex structure analysis • 3D modeling • Geological map reading • Geophysical profile interpretation	• Tectonics, magmatism • Underground geometry • Geological cartography • Geophysical prospecting	• Interprets geophysical data • Reconstructs geometries • Reads geological symbols • Models 3D structures
Temporal Vision	• Radioactive waste management • Long-term planning	• Geological stability • Future responsibility	• Evaluates over 100,000 years • Considers uncertainties
Environmental Awareness	• Anthropocene analysis • Ecosystem services	• Human geological impacts • Georesource value	• Quantifies impacts • Evaluates sustainability
Communication	• Information campaigns • Dissemination materials	• Seismic risk • Prevention	• Adapts to public • Balances precision-clarity • Measures impact

ASSESSMENT TOOLS

Level-based Rubrics

Primary: 🟢 Autonomous - 🟡 Guided - 🔴 Initial

Secondary I: 🟢 Competent - 🟡 Developing - 🔴 Basic

Secondary II: 🟢 Expert - 🟡 Competent - 🔴 Developing

Geological Portfolio

- Field observation notebook
- Sample collection with data sheets
- Conceptual maps of evolution
- Metacognitive reflections

Observable Indicators

- **Systemic**: uses "depends on", "is connected", draws arrows between elements
- **Critical**: asks "How do they know?", compares sources, verifies information
- **Communicative**: adapts language, uses analogies, verifies comprehension

15.13 Vertical progression of skills: from concrete to abstract

An effective vertical geoscience curriculum should accompany students' cognitive development, presenting content and activities appropriate for each level while maintaining a coherent vision of the skills to be developed.

Primary school (6-10): Focus on direct observation, manipulation, and sensory exploration of rocks, minerals, and immediately accessible landscape features. Activities should be rooted in everyday experience and the local context.

Middle school (10-14): Introduction of more abstract concepts, first generalizations, use of symbols and representations. Students can begin to understand processes that are not directly observable through models and simulations.

Upper secondary school (14-18) Development of formal thinking, understanding of complex systems, use of sophisticated mathematical and conceptual simulations. Students can tackle abstract concepts such as deep geological time, nonlinear dynamics in Earth systems, and complex interactions between the geosphere and other systems.

An effective vertical curriculum should gradually expand spatial and temporal horizons, from the local and present experience observable in primary school, to the regional and historical dimension in lower secondary school, to a global view of geological processes and an awareness of deep geological time in upper secondary school.

The skills developed through geosciences constitute “cognitive tools” that enrich thinking and action in any field of knowledge. Competence-oriented teaching becomes a powerful vehicle for comprehensive education, preparing students to face complex challenges with robust and flexible conceptual tools. The ability to think systemically, analyse complex relationships, reason on multiple scales, and communicate effectively is a valuable intellectual asset for facing the challenges of a rapidly changing world.

“Xuzhuang” formation from the early Cambrian period in the North China Block.

Further reading

Geologically - Ideas for teaching www.geologicamente.it

Mogk, D.W., & Goodwin, C. (2012). *Learning in the field: Synthesis of research on thinking and learning in the geosciences*. In: Earth and Mind II: A Synthesis of Research on Thinking and Learning in the Geosciences (pp. 131-163). Geological Society of America.

Research on the development of critical thinking and investigative skills through the **European Council Recommendation** (2018). *On key competences for lifelong learning*. Official Journal of the European Union.

CHAPTER 16.
BEYOND THE HAMMER AND COMPASS: THE ROLE OF DIGITAL TECHNOLOGIES IN GEOSCIENCE EDUCATION

Geosciences are rapidly evolving beyond the traditional image of the field geologist with a hammer and compass. While field observation remains fundamental, new technologies are radically transforming the way we study and understand Earth systems. While chemistry, biology, and physics offer extensive laboratory opportunities, geosciences face unique challenges in providing practical experiences to students, who have traditionally had to rely on field trips, rock sample handling, and map study.

Teaching geosciences requires a multimodal approach that allows them to understand complex processes operating on extremely diverse spatial and temporal scales. In this context, the advent of digital technologies has significantly expanded educational possibilities, offering new ways to visualize, analyse, and interact with geological phenomena.

However, there is a significant gap in the availability of digital tools specifically designed for teaching geosciences: most educational applications are developed for university students or professionals, creating a void in secondary education, precisely where these tools would be particularly useful for introducing the basic concepts of the discipline and stimulating students' interest.

16.1 The digital divide in geosciences and opportunities for innovation

Unlike other scientific disciplines, geosciences have traditionally struggled to effectively integrate digital tools into pre-university teaching. This lag manifests itself in several ways:

1. **Lack of age-specific applications**: while there are numerous educational applications for mathematics and physics suitable for youngers, digital tools for geosciences are often designed for advanced users.

2. **Excessive complexity**: Many professional geological software programs have steep learning curves, making them inaccessible to students and teachers without specialized training.
3. **Limited curriculum integration**: There is a lack of clear guidelines on how to effectively integrate digital tools into existing Earth science curricula.
4. **Perception of the discipline**: The persistent image of geosciences as a "field" discipline has slowed the adoption of innovative digital approaches.

Paradoxically, this lag persists despite students being 'digital natives' and that the minerals studied in geosciences are essential components of the technologies they use every day.

Enrolment data for Earth Science courses show contrasting trends that warrant further analysis. There is a general decline in enrolment in traditional geology courses, while there is significant growth in interdisciplinary programs that integrate geosciences with digital, communication, or environmental management skills.

This phenomenon suggests not so much a lack of interest in the discipline itself, but rather an evolution in expectations regarding professional skills. Young people are looking for educational paths that integrate traditional geological knowledge with cross-cutting skills relevant to the contemporary job market.

16.2 Digital Technologies as Tools for Educational Transformation

Today's digital technologies offer unprecedented opportunities to transform how students engage with geosciences. The solution lies in implementing digital tools that allow them to experiment with geological phenomena in an interactive and meaningful way. All applications and digital tools presented in this chapter are available in free versions suitable for school use, ensuring accessibility to all schools regardless of available budget.

Additionally:

- All tools work on standard devices (smartphone/tablet/PC)
- Only tools that respect privacy have been considered
- Priority has been given to intuitive interfaces usable without advanced technical training

Geographic Information Systems (GIS) and Geospatial Technologies

Geographic and cartographic skills, far from being supplanted by modern technology, must be enhanced and strengthened through digital tools. These skills require an active approach where students engage in observation, investigation, and analysis, rather than passively receiving information.

Digital surveying and visualization, for upper-year students or technical tracks, have revolutionized how they interact with geographic and geological data:

- **Google Earth Pro** serves as the most known platform for geospatial analysis, allowing exploration of Earth from multiple perspectives. Combined with satellite imagery and LiDAR, it enables observation of environmental changes at scales previously impossible in educational settings, helping understand large-scale geological processes and their impacts. Its intuitive interface makes it ideal for introducing basic concepts of geomorphology and territorial analysis, even for pupils.
- **QGIS**, an open-source geographic information system, offers more advanced functionality for territorial analysis; it facilitates the creation of topographic maps, terrain analysis, slope calculations, and geological data visualization.

Useful online resources also include:

- **OpenStreetMap**: https://www.openstreetmap.org/
- **NASA Worldview**: https://worldview.earthdata.nasa.gov/
- **Natural Earth**: https://www.naturalearthdata.com/

Structural Analysis: Understanding the Third Dimension

Structural geology generally falls outside ministerial programs, but observation of geological sites or particular outcrops allows direct observation of tectonic processes that have shaped Earth's crust over millions of years. For example, the **Dolomites** offer examples of horizontal stratification and folds, where ancient Triassic seafloors were uplifted and deformed during Alpine orogenesis. The **Grand Canyon** exposes one of the world's most complete stratigraphic sections, where each layer tells a chapter of North American geological history.

In **Iceland**, along the **Thingvellir Fault**, one can literally walk on the Mid-Atlantic Ridge, observing how Earth expands creating new oceanic crust. These outcrops make plate tectonics perceptible, allowing one to "touch" the planet's geological history.

Similar world-class geological exposures exist globally: the Zagros Mountains in Iran displaying spectacular fold structures, the Appalachian Valley and Ridge Province showing ancient mountain-building processes, the Flinders Ranges in Australia revealing billion-year-old geological history, or the Scottish Highlands preserving evidence of ancient continental collisions. Each region offers distinctive opportunities to observe tectonic processes recorded in rock.

Digital resources offer powerful tools for visualizing and analysing some of these complex geological structures:

- **Outcrop 3D** presents detailed three-dimensional models of real outcrops through Structure from Motion (SfM) techniques, using series of photographs taken from different angles. The resulting 3D sample allow virtual exploration of significant geological sites, precise measurements of geological structures, and analysis of complex spatial relationships.

Discontinuities on Portuguese coasts

- **Visible Geology** (https://www.visiblegeology.com) and especially the Geology Explorer option represents an innovative tool that enables creation and manipulation of three-dimensional geological models, visualizing layers, folds, faults, and other geological structures. It's possible to modify parameters like inclination, direction, layer thickness and immediately observe how these changes influence the geometry of resulting structures. The site's educational section offers guided exercises connecting theoretical concepts to practical visualizations, making this resource particularly valuable for distance learning or as a complement to traditional lessons.

Representation created by students of the Portuguese outcrop

Professional geological societies worldwide maintain educational resource repositories.

The Italian Geological Society (https://www.socgeol.it/N2629/risorse-di-didattica-online.html), the Geological Society of America (https://www.geosociety.org/education), the Geological Society of London (https://www.geolsoc.org.uk/education), and similar organisations in other countries provide curated collections of teaching resources, many freely accessible to educators globally.

Exploring Rocks and Minerals: From Analog to Digital

The study of rocks and minerals, often considered less engaging than other aspects of geosciences, can be revitalized using dedicated mobile applications.

Platforms like Sketchfab host numerous examples like those from the University of Turin

https://sketchfab.com/3dmodels/p5granitobiancobiotitico1bb7762d90844f27aecd2491003c9 a58,

creating a virtual library of educational materials accessible from anywhere. These tools transform abstract concepts of structural geology into manipulable visual representations, making concepts accessible that traditionally prove difficult to visualize through textbooks alone.

National geological surveys and research institutions in many countries offer educational programs that bring geological expertise directly to classrooms. In Italy, ISPRA (Higher Institute for Environmental Protection and Research) offers the 'Educational Project Rock and Mineral Laboratory,' with educational seminars conducted by expert technicians using microscopes and specimens from their collections. Similar programs exist through the USGS in the United States, the British Geological Survey, Geoscience Australia, the Geological Survey of Canada, and other national institutions worldwide, each offering regionally specific educational resources and outreach programs that connect students with professional geological expertise and collections.

Rock identification apps like Rock Identifier represent an accessible starting point, allowing photography of samples and obtaining immediate identifications accompanied by detailed information on chemical composition, hardness, and geological origin. These applications transform smartphones into scientific instruments, enabling rapid identifications both in class and during field trips. The basic version includes a database of thousands of rocks and minerals.

The Rockd app transforms smartphones into powerful tools for geological field exploration. This application offers access to numerous interactive geological maps and allows recording observations even offline. The app enables visualization of surrounding territory geology, recording observations with GPS coordinates, accessing information on local rock formations, and contributing to a global scientific database. The app also includes paleogeographic maps showing how Earth appeared in past epochs, proving specifically useful for preparing educational excursions and systematically documenting observations during outings.

Simulating Volcanic Activity: Bringing Volcanoes to Class

Volcanic processes, traditionally difficult to study directly, can be made more accessible through digital tools, like those from the Smithsonian https://volcano.si.edu/E3/

Volcano monitoring institutions worldwide provide real-time data valuable for educational purposes. Italy's INGV (National Institute of Geophysics and Volcanology) constantly monitors Etna, Stromboli, Vesuvius, and other Italian volcanoes, providing webcams and data streams. Similarly, the USGS Volcano Hazards Program monitors volcanoes throughout the United States and globally, the Japan Meteorological Agency tracks activity across the Pacific Ring of Fire, GeoNet monitors New Zealand's volcanic systems, and INGV-OV observes Latin American volcanoes. These institutional resources provide students worldwide with access to real-time volcanic monitoring data from diverse tectonic settings.

The University of Rhode Island's lava flow simulator offers an interactive experience directly from the browser, allowing modification of parameters like magma type and eruption velocity, observing how these factors influence lava flow behaviour and associated risks. This tool makes possible exploration of "what-if" scenarios, helping understand why some volcanoes are more dangerous than others and how topographic characteristics influence flow direction.

For Android devices, the "volcanoSimulator" app is available, offering basic simulations of lava movement. Simpler than professional web simulators, it's accessible and can stimulate student interest. Teachers should use these tools as starting points for deeper discussions about volcanic processes, always integrating with real data from historical eruptions and satellite images. The combination of digital simulations with real case studies allows development of more complete and nuanced comprehension of volcanic phenomena.

Digital Palaeontology: exploring past life

The Paleobiology Database represents an extraordinary resource for paleontological education. This public database, maintained by scientists from 24 different countries, offers access to information on millions of fossils and enables sophisticated research on biodiversity, extinctions, and faunal changes in geological time. One can conduct autonomous research on specific geological periods, compare biodiversity of different eras, study mass extinction events, and create graphs of life evolution on Earth.

Platforms like Sketchfab and Morphosource host thousands of 3D fossil models created through scans of real samples. These objects can be rotated, enlarged, and in some cases, virtually "disassembled" to study internal anatomy. For schools with 3D printers, many of these examples can be downloaded and printed to create tactile fossil collections, particularly useful for students with learning difficulties or visual impairments.

TimeTree is an application available both as a website and mobile app that enables exploration of evolutionary relationships between all living and fossil organisms. The app enables discovery of when evolutionary lines of different organisms separated, comparison of evolution times of different groups, and understanding the interconnection of all life on Earth. The simple interface provides immediate answers to questions like "when did dogs and cats evolutionarily separate?" stimulating curiosity and discussions about life's history.

Practical implementation in schools

Effective integration of digital technologies in geoscience teaching requires careful planning and a gradual approach. Before introducing new technologies, it's fundamental to evaluate available resources in the school. Many recommended applications work on standard smartphones and tablets, making them accessible even in contexts with limited budgets.

The success of technological integration depends mainly on educator preparation. It's advisable to start with one or two simple applications, allowing teachers to gain familiarity before introducing more complex tools. Many apps include integrated tutorials and support communities that facilitate learning. Platforms like ESRI Italy's School of Where offer courses for teachers, while many geological software developers provide educational webinars specific for school use.

- **ArcGIS and the School of Where** (ESRI Italy): The ArcGIS platform offers a complete ecosystem of tools for geospatial analysis and visualization particularly suited to educational context. The "School of Where" initiative https://resources.esriitalia.it/37-formazione/583-the-school-of-where provides specific resources for teaching geosciences

through geospatial technologies, with programs dedicated to schools of all levels.
- **Story Maps** represent an innovative tool for integrating geological data into engaging narratives. Platforms like https://resources.esriitalia.it/34-prodotti/piattaforma-esri/368-story-map-dai-potere-alle-tue-storie allow creation of narrative paths that integrate interactive maps, images, videos, and text, transforming geoscience learning into an immersive experience.
- **GeoInquiries** https://www.esri.com/itit/industries/k12education/geoinquiries/world-geography offer Inquiry-Based educational methodologies, where real geospatial data is used to explore geological phenomena, simultaneously developing analytical skills and critical thinking.

The goal remains to enhance, not replace, traditional learning based on direct observation. Tech resources are most effective when they prepare, support, and deepen real experiences with natural phenomena.

New Frontiers: Augmented Reality, Virtual Reality, and 3D Printing

Augmented reality (AR) applications are beginning to transform geological learning experience, allowing overlay of digital information onto the real world and enabling "seeing" geological structures hidden beneath the surface during field excursions. Although still in development, some apps already allow visualization of virtual geological sections by pointing smartphones at rock outcrops, helping understand three-dimensional subsurface structure and geological formation continuity.

VR systems can virtually transport to otherwise inaccessible geological environments: from ocean depths to active volcanic calderas, from Antarctic glaciers to deep mines. Many schools are investing in economical VR headsets that work with smartphones. VR content for geosciences is still limited but rapidly expanding. Google Earth VR already enables immersive geological explorations of famous formations like the Grand Canyon, while some museums offer virtual visits to paleontological collections.

Many schools now have 3D printers, opening extraordinary possibilities for geoscience teaching. Physical replicas of fossils, crystals, and geological formations can be printed to create personalized tactile collections. Teachers can produce representations of rare fossils normally inaccessible or volcanoes with different morphologies to understand the relationship between eruption type and shape, as well as three-dimensional geological sections to visualize stratigraphy.

3D printing enables projects combining geosciences, technology, and art, producing learning experiences that integrate creativity and scientific rigor.

Emerging technologies also enable geographically distant students to collaborate in shared virtual environments. These digital spaces can host interactive geological representations, allowing twin classes to explore virtual rock formations together or collaborate on paleontological research projects, opening new possibilities for global geological education.

Citizen science projects and connection with scientific community

Digital technologies enable active contribution to scientific research through citizen science projects. Rockd allows sharing geological observations with a global community of researchers, while interdisciplinary projects like eBird (for biodiversity) or iNaturalist can be used for studies combining geology and biology. These projects offer opportunity to understand how scientific research actually works, experiencing the importance of systematic data collection and international collaboration.

Digital tools also offer new ways to assess learning. Geolocated interactive quizzes can test ability to identify rock formations during excursions, while digital portfolios allow documenting progress in mineral identification or understanding of geological processes. Gamification elements like leaderboards, badges, and challenges increase engagement, transforming rock identification or fossil discovery into engaging and competitive activities.

Implementation considerations and future perspectives

When using mobile apps and online platforms, it's important to consider privacy, as many applications collect location data and personal information. Schools must verify privacy policies and, when necessary, request appropriate consents. Use of technology must also be balanced with environmental considerations: encouraging virtual rather than physical collection of geological specimens can contribute to environmental conservation, but it's important that students also understand the importance of physical conservation of geological ecosystems.

Schools wishing to invest in geoscience technologies should prioritize versatile and low-maintenance tools. Smartphones and tablets with free or low-cost apps offer the best cost-benefit ratio, while investments in more sophisticated hardware should be carefully planned based on real curricular needs. Teachers must be supported with continuous training opportunities, as technological evolution is rapid and new tools emerge constantly.

Digital skills developed through use of geoscientific technologies are transferable and valuable for future work. The ability to use scientific databases, interpret 3D visualizations, and work with geospatial data are increasingly demanded in many professional sectors, making this educational investment exceptionally forward-thinking.

16.3 Towards Integrated School Geoscience

Effective integration of digital technologies in geoscience teaching represents a natural evolution responding to contemporary student needs. These tools, when used appropriately, can transform geoscience learning from a traditionally perceived static subject to a dynamic and engaging discipline.

The success of this integration depends on educators' ability to balance innovation and tradition, using technologies to amplify rather than replace direct experience with natural phenomena. The geologist of the future will certainly continue to use hammer and compass, but will also have the possibility to document, analyse, and share discoveries through digital tools that will enormously amplify their capacity to observe and understand our planet's complexity.

Investment in these technologies, accompanied by adequate teacher training and curricular integration, can transform geological education, preparing a new generation of scientifically literate citizens capable of facing future environmental challenges.

The Peuterey ridge of Mont Blanc and the retreat of the Brenva glacier

Further reading

Bouziat, A., Schmitz, J., Deschamps, R., & Labat, K. (2020). Digital transformation and geoscience education: New tools to learn, new skills to grow. *European Geologist*, 50. DOI: 10.5281/zenodo.4311379

Gunderson, K.K., et al. (2020). *Recent digital technology trends in geoscience teaching and practice*. GSA Today, 30(3), 4-11. DOI: 10.1130/GSATG404G W.1

McCaffrey, K.J.W., et al. (2005). *Unlocking the spatial dimension: digital technologies and the future of geoscience fieldwork*. Journal of the Geological Society, 162(6), 927-938.

De Paor, D.G. (2016). *Virtual rocks*. GSA Today, 26(8), 4-11. DOI: 10.1130/GSATG257A.1

Mašterová, V. (2023). *Learning and teaching through inquiry with geospatial technologies: A systematic review*. European Journal of Geography, 14(3), 42-54. DOI: 10.48088/ejg.v.mas.14.3.042.054

CHAPTER 17.
GEOSCIENCES AND THE SUSTAINABLE DEVELOPMENT GOALS

In the context of growing global environmental challenges, geosciences play a fundamental role in addressing these pressing issues of our time. The Sustainable Development Goals (SDGs) of the United Nations 2030 Agenda represent the global roadmap for creating a more sustainable and equitable future, and geosciences are intimately connected to most of these goals.

The centrality of geosciences in the SDGs is widely recognised by the international scientific community, although it remains insufficiently valued in educational and decision-making contexts globally. This is not a marginal or ancillary role: geosciences represent the knowledge backbone necessary to understand, address, and solve many of the challenges outlined in the 2030 Agenda. This centrality stems from the very nature of geosciences, which study the Earth system in its complexity and its interactions with human activities—a reality that transcends national boundaries and educational systems, making geoscience literacy a universal necessity for sustainable development.

The contribution of geosciences to the SDGs is multifaceted: from understanding the natural processes that regulate our planet, to identifying and sustainably managing resources, to mitigating natural risks and designing strategies for adapting to global change. Geosciences provide the conceptual and methodological tools to analyse the complex interdependencies between the geosphere, hydrosphere, atmosphere, and biosphere, offering a systemic view that is indispensable for any sustainability strategy.

Paradoxically, despite this crucial importance, geosciences often remain in the background of discussions on sustainable development, which tend to favor economic, sociological, or engineering approaches. This marginalization represents not only a limitation to the global knowledge of environmental issues, but also a concrete obstacle to the implementation of effective and lasting solutions.

17.1 Fundamental Environmental SDGs

Clean Water and Sanitation (SDG 6)

Geologists play a crucial role in sustainable water management that goes far beyond traditional perceptions of the profession. Their expertise in hydrogeology and water resource management is essential to ensuring sustainable access to clean water. Despite this importance, the current curriculum structure significantly underutilizes opportunities for practical learning in this area.

Key topics such as hydrogeological cycles, aquifer systems, and groundwater dynamics should be core components of the curriculum, supported by comprehensive practical activities that include:

· Field studies of local aquifer systems

· Water quality testing and monitoring practices

· Analysis of local water management cases

These practical components are frequently reduced to abstract theoretical concepts in many educational systems worldwide. This creates a significant disconnection from real-world experience and preventing students from fully grasping the social relevance of geological knowledge in the context of sustainable water management. This gap between theory and practice represents a global challenge in geoscience education, as water security issues affect communities across diverse geographical and socioeconomic contexts, from arid regions facing water scarcity to coastal areas dealing with saltwater intrusion, from developed nations managing aging water infrastructure to developing countries establishing new water systems.

Clean and Affordable Energy (SDG 7)

Geologists are at the forefront of the transition to renewable energy, with a particularly crucial role in the development of geothermal energy across its entire spectrum.

In many educational contexts worldwide, there is a tendency to focus primarily on high-enthalpy geothermal systems, such as those found in tectonically active regions like Iceland, New Zealand, the Philippines, or the Larderello field in Tuscany, Italy, neglecting the vast potential of medium- and low-enthalpy geothermal applications. These medium- and low-enthalpy systems, which can be implemented in a much wider range of geological contexts—from the stable cratons of continental interiors to sedimentary basins worldwide—represent significant opportunities for sustainable heating and cooling solutions applicable in both urban and rural environments across diverse climate zones.

The curriculum should be enriched by exploring in depth how geological expertise guides site assessment for various renewable energy infrastructure projects, from large-scale power plants to smaller domestic installations.

A more comprehensive educational approach in this area would help students understand the versatility of geothermal applications and the diverse career opportunities available in this rapidly growing sector. The energy transition needed to address the climate crisis will require a growing number of geothermal experts, making this gap problematic for the development of future professional skills appropriate to the challenges of energy sustainability.

Fighting climate change (SDG 13)

Geologists offer unique insights through their study of the Earth's past climate records, providing crucial context for recognise current climate change and its future implications. The geological record serves as a valuable archive of past climate variations, offering evidence of both gradual changes and abrupt climate upheavals, which have profound implications for our awareness of contemporary climate dynamics.

According to IPCC data, the rate of current climate change far exceeds anything observed in the geological record over the past 800,000 years, a fact that underscores the urgency of the current situation. The curriculum can and should demonstrate how geological knowledge informs our understanding of and adaptation strategies to today's climate change, linking past Earth processes to present-day environmental challenges in a coherent and meaningful way.

Relevant is the study of extreme weather events in the geological record and their impact on terrestrial landscapes. Geological evidence of ancient floods, prolonged droughts, and rapid temperature changes provides crucial information about the potential impacts of the climate change we are experiencing. The role of geologists in climate action extends beyond merely recognising past climates to include practical applications in adaptation and mitigation, including the assessment of geological risks exacerbated by climate change.

Life on Earth (SDG 15)

Geologists' knowledge of soil systems and terrestrial processes is crucial for the conservation of ecosystems and biodiversity. Despite this centrality, the treatment of soil systems and terrestrial processes in curricula often remains superficial, characterized by limited practical activities that prevent a deep understanding of ecosystem interconnections. Essential practical components that should be included in educational programs include soil sampling and analysis techniques, field mapping of different soil types, methods for assessing ecosystem health, and laboratory studies of soil composition.

This predominantly theoretical approach is not unique to any single educational system but represents a widespread challenge in geoscience education globally, whether in resource-rich developed nations or in developing countries where direct knowledge of soil-ecosystem relationships might be even more critical for local food security and land management. The disconnect affects students' ability to understand how geological processes maintain biodiversity and provide essential ecosystem services across diverse biomes. These range from tropical soils supporting rainforest ecosystems to permafrost regions experiencing rapid change, from agricultural heartlands requiring sustainable soil management to fragile mountain ecosystems balancing between erosion and vegetation.

These practical activities are essential to recognise the interconnections between geological processes and ecosystem health but are frequently neglected in favour of theoretical instruction, preventing students from developing an experiential understanding of how soils function as a critical interface between the lithosphere, biosphere, hydrosphere, and atmosphere."

Life below water (SDG 14)

Marine geosciences are central to the protection of aquatic ecosystems, providing essential knowledge about coastal sedimentary processes, marine habitat dynamics, and the impacts of human activities on aquatic environments. The study of coastal geomorphology and land-sea interactions provides tools for the protection of marine biodiversity and the sustainable management of coastal resources.

The curriculum should integrate elements of marine geology, coastal sedimentology, and hydrogeology of land-sea transition zones, with a focus on the interactions between geological and biological systems in aquatic environments. Practical activities should include coastal sediment analysis, marine habitat mapping, and assessment of the impacts of sea level rise on coastal ecosystems, elements that are often neglected in traditional geological training.

Understanding geological processes in marine environments is particularly crucial in the context of climate change, as changes in ocean circulation, water acidification, and alterations in marine biogeochemical cycles have profound implications for global marine ecosystems and, consequently, for the human communities that depend on them.

These limitations in practical and laboratory activities in all these areas not only negatively affect students' conceptual comprehension, but also impact their career choices and, ultimately, our collective ability to achieve sustainable development goals. The priority given to theoretical knowledge over practical experience creates a significant gap between formal education and the skills needed to address real-world environmental challenges. This disconnect specially affects the ability to understand the practical applications of geological knowledge and their potential role in addressing urgent ecological and conservation challenges.

17.2 Urban Development and Infrastructure

Industry, Innovation, and Infrastructure (SDG 9)

In the context of Industry, Innovation, and Infrastructure, geoscientific expertise should be recognised as indispensable for sustainable urban planning, infrastructure development, and natural risk mitigation. However, the current secondary school curricula are significantly lacking in addressing these crucial aspects of geoscience application.

Although the curriculum nominally includes concepts of geological hazards and land-use planning, it would need substantial expansion to adequately cover how geologists assess ground stability for infrastructure projects and how this knowledge influences critical design and construction decisions.

Students in educational systems worldwide should have the opportunity to learn through real-world examples from their local area, understanding how geological knowledge informs urban development and infrastructure planning decisions. This connection between theory and practice—whether examining earthquake-resistant design in seismically active regions like Japan, California, or the Mediterranean, analysing subsidence issues in cities built on soft sediments like Bangkok, Venice, or Mexico City, or perceiving groundwater management beneath rapidly expanding urban centres from Jakarta to São Paulo—would allow globally to appreciate the immediate relevance of geology in creating safe and sustainable urban environments within their own geographical and cultural contexts.

Sustainable Cities and Communities (SDG 11)

Despite the critical role that geologists should play in urban development and safety, the treatment of geological hazards and land-use planning in the curriculum remains largely theoretical and disconnected from real-world applications. Urban geology studies, which should be a cornerstone of modern geoscience education, are often completely absent from the curriculum. Local examples of how geological knowledge shapes urban planning, which would be invaluable for students' knowledge, are rarely, if ever, incorporated into classroom learning.

This educational gap persists across diverse urban contexts worldwide, from earthquake-prone cities requiring seismic hazard assessment to coastal metropolises facing subsidence and sea-level rise, from rapidly expanding megacities in the developing world to established urban centres in developed nations requiring infrastructure renewal.

Students should have the opportunity to engage with practical case studies from their own cities, understanding how groundwater systems influence urban development and how geological factors determine fundamental construction and planning decisions. They should explore how seismic resistance, soil stability, and other geological factors influence urban design and land use planning.

These crucial connections between geology and urban development are typically overlooked or only superficially addressed in educational systems worldwide, limiting the awareness of the essential role of geosciences in creating safe, resilient, and sustainable cities—whether they live in Tokyo, London, Shanghai, Lagos, or any other urban centre facing the complex interplay between geological conditions and human settlement.

Responsible Consumption and Production (SDG 12)

Geologists play an essential role in promoting sustainable consumption and production patterns through their knowledge of Earth's resource cycles and planetary boundaries. Knowledge of mineral formation processes, resource regeneration rates, and biogeochemical cycles allows for the establishment of sustainable exploitation thresholds and the design of circular economy systems.

Geoscience education should emphasize the principles of resource economics, from assessing the ecological footprint of materials to designing product life cycles that minimize waste. Students should be exposed to life cycle assessment techniques and resource sustainability analysis, developing practical skills in identifying sustainable alternatives and valorising geological heritage from a circular perspective.

The transition to a circular economy requires a deep understanding of the Earth's material cycles, making geosciences central to sustainable production and consumption strategies. This dimension should be explicitly integrated into educational curricula, linking geological principles to the everyday challenges of material sustainability.

17.3 Socioeconomic Impacts

No Poverty (SDG 1) and Zero Hunger (SDG 2).

Geological expertise plays a vital role in addressing the key social challenges of poverty reduction and food security through multiple interconnected pathways. Geological knowledge of soil systems directly supports agricultural sustainability by providing the basis for land management practices that maximize productivity while maintaining long-term soil health. At the same time, responsible management of mineral resources, guided by sound geological expertise, can contribute significantly to local and national economic development by creating employment opportunities and generating resources for social investment.

Another critical aspect is the assessment and mitigation of geological hazards, which helps protect vulnerable communities and their livelihoods from catastrophic events such as landslides, floods, and earthquakes, events that can quickly push entire communities into poverty.

Current curricula in many countries often overlook these fundamental connections, depriving students of an understanding of how geological expertise concretely supports sustainable agricultural practices through soil analysis and pedological studies across diverse agricultural contexts—from industrial farming in temperate zones to subsistence agriculture in tropical regions, from irrigated lands dependent on aquifer management to rain-fed systems vulnerable to climate variability. Similarly, students worldwide rarely explore how responsible mineral resource management and effective natural hazard preparedness contribute to community resilience and economic development, whether in mineral-rich developing nations seeking to avoid resource curse dynamics or in developed countries managing the sustainable extraction of strategic materials.

Health and Well-being (SDG 3)

The goal of good health intersects with geology through the emerging field of medical geology and environmental health assessment, a growing interdisciplinary area. This connection is particularly evident in regions with a rich hydrogeological heritage, where natural mineral waters and their implications for human health have been historically significant and continue to represent an important intersection between geology and medicine.

The curriculum in many educational systems could effectively demonstrate these relationships through the integration of practical environmental contamination assessment studies, such as the analysis of heavy metals in soils and water, and through radon monitoring programs in buildings—approaches applicable across diverse geological settings from granitic regions with naturally elevated radon levels to industrial areas with legacy contamination issues. Students globally could explore how the geological composition of an area influences drinking water quality, the presence of essential or potentially toxic minerals in the food chain, and even the geographical distribution of specific diseases related to geological factors, such as fluorosis in regions with fluorine-rich groundwater, iodine deficiency in mountainous areas far from oceanic iodine sources, or arsenic poisoning in regions with naturally arsenic-rich aquifers like parts of Bangladesh, Argentina, or the western United States.

Integrating these topics into the curriculum would not only enrich the comprehension of the connections between geosciences and human health but could also inspire interdisciplinary careers at the intersection of geology, public health, and environmental medicine, a rapidly expanding field that will require professionals with cross-disciplinary skills to address emerging global health challenges related to environmental factors.

Decent work and economic growth (SDG 8)

Perhaps less obvious but equally significant is the potential role of geosciences in promoting sustainable and inclusive economic growth. Geological knowledge should support the responsible extraction and management of natural resources, a sector that, if properly developed and regulated, can create sustainable job opportunities, particularly in developing economies where natural resources often represent an important source of national wealth.

When properly managed with geological expertise and attention to environmental and social impacts, mining and extractive industries can serve as powerful engines of economic development while maintaining environmental responsibility and respect for local communities. This would be specifically relevant in the current context, given the growing demand for strategic minerals essential for green technologies such as batteries, solar panels, and wind turbines, whose sustainable extraction requires in-depth geological knowledge.

Ethical sourcing of materials, which considers not only the environmental but also the social aspects of mining, is another area where geological expertise intersects with economic and human rights considerations.

Unfortunately, these contemporary and socially relevant applications of geosciences rarely find their way into standard curricula in educational systems worldwide, limiting perception of the economic value of geosciences and their potential contribution to sustainable and socially responsible economic development. This educational gap affects students across diverse economic contexts—from mineral-rich developing nations where responsible resource extraction could drive economic transformation, to developed economies requiring professionals skilled in sustainable resource management, to regions experiencing economic transitions where geotourism and sustainable resource enhancement offer alternative development pathways

17.4 Educational and Social Progress

Quality Education (SDG 4) and Gender Equality (SDG 5)

Quality Education and Gender Equality naturally converge within the geosciences through an integrated educational approach that promotes both scientific excellence and inclusive participation. The geoscience curriculum has the unique potential to achieve these dual goals by emphasizing hands-on field experiences and direct scientific investigation methods, thereby creating learning opportunities that transcend traditional gender barriers present in some science education contexts.

Field-based learning, characteristic of quality geological education, creates a democratizing environment in which all students, regardless of gender, can develop practical skills, scientific confidence, and investigative autonomy. This experiential approach reduces the gender gap in science by valuing multiple forms of intelligence and learning styles, creating a more inclusive educational environment than traditional theoretical learning contexts.

A well-designed curriculum should also actively highlight the diverse contributions of geoscientists across gender and cultural boundaries, showcasing the achievements of women geologists alongside their historical and contemporary contributions to the field. This not only provides diverse role models, but also concretely demonstrates the discipline's evolution toward greater inclusivity. Presenting geology as a science in which people of all genders have made and continue to make significant contributions can encourage broader and more diverse participation.

This integrated educational framework transforms abstract geological concepts into tangible components of global sustainability, helping students develop both the practical skills and inclusive mindset necessary to address future environmental and social challenges. The holistic approach to geology education, which values both scientific rigor and diversity of perspectives, prepares them not only to be competent geoscience professionals, but also to be informed global citizens and promoters of equity in the scientific community and beyond.

Peace, Justice, and Strong Institutions (SDG 16)

The link between geosciences and peace may seem less obvious, but shared management of natural resources and cross-border cooperation in geological risk mitigation are powerful tools for peacebuilding. Water diplomacy, in particular, demonstrates how hydrogeological knowledge can facilitate cooperation agreements even between countries with complex political relations.

Geoscience education should include case studies on shared governance of natural resources, environmental conflict resolution, and international scientific cooperation in monitoring transboundary geological risks. Exploring these political and diplomatic dimensions of geoscience would enrich students' education, preparing them to contribute to stronger institutions and more inclusive decision-making processes on environmental issues.

Reducing Inequalities (SDG 10)

Geosciences contribute to reducing inequalities through equitable access to natural resources and shared management of geological risks. Proper mapping and assessment of natural resources can guide more equitable distribution policies, especially in mineral-rich but economically disadvantaged regions. The geoscientific approach to environmental justice highlights how the most vulnerable communities are often disproportionately exposed to geological risks and the negative impacts of resource extraction.

The geoscience curriculum should include these ethical and social justice dimensions, exploring how geological knowledge can be used to identify and correct inequalities in exposure to natural hazards, access to water resources, and the distribution of negative environmental impacts. The integration of case studies involving different communities would allow students to understand how decisions based on geological data can promote or hinder social equity.

Geological knowledge can also provide economic opportunities in disadvantaged regions through the identification of sustainable local resources and the promotion of geotourism, thus contributing to the reduction of territorial inequalities. Furthermore, democratizing access to geological data through open-source platforms and accessible technologies can reduce information disparities that often penalize communities with fewer economic resources in environmental decision-making.

Students should be made aware of how methodological choices in geological research can have different social implications, developing critical awareness of the dimensions of equity and environmental justice that underlie the professional practice of geosciences. This perspective is essential for training professionals who can consciously contribute to reducing inequalities through their technical expertise.

17.5 Partnerships for the Goals (SDG 17) – Recommendations

Geosciences thrive through interdisciplinary and international collaborations, offering a model of scientific partnership that reflects the spirit of SDG 17. Global seismic monitoring networks, international ocean drilling programs, and collaborations for the study of the Earth system demonstrate the effectiveness of transnational scientific cooperation.

Highlighting these collaborative dimensions of geosciences, showing how sharing data, methodologies, and research infrastructure accelerates scientific and technological progress. Students should be exposed to citizen science projects and North-South cooperation initiatives in geosciences, developing international collaboration skills essential for addressing the global challenges of the future.

The spirit of partnership that characterizes best practices in geosciences should also guide the rethinking of geoscience education for sustainable development.

These recommendations apply across diverse educational contexts, recognising that implementation strategies will necessarily vary based on regional resources, institutional frameworks, and local educational priorities, while maintaining the universal goal of integrating sustainability into geoscience education.

To fully realize the potential of geosciences in the 2030 Agenda, we recommend:

1. **Integrated curriculum review** that deliberately intertwines geological concepts with the SDGs, showing the direct connections between geological theory and contemporary challenges through partnerships between educators and professionals.
2. **Enhanced practical training** through collaborations with research organisations, companies, and public institutions that can offer real-world contexts for applying geosciences to sustainability.
3. **Networked teacher training**, creating transnational communities of practice that share innovative teaching methodologies to integrate the SDGs into geoscience education.
4. **Interdisciplinary collaborations** that transcend traditional boundaries, developing educational approaches that reflect the interconnected nature of sustainability challenges.
5. **Collaborative digital platforms** that connect geoscience students around the world, enabling the exchange of experiences and the co-creation of solutions to global problems.
6. **Community engagement** through participatory approaches that value traditional and local knowledge and integrate it with scientific knowledge.

7. **Transnational evaluation networks** that measure the effectiveness of educational programs and share best practices across geographical and disciplinary boundaries.

The centrality of geosciences in the 2030 Agenda is not only a matter of disciplinary relevance but represents a strategic necessity for the training of professionals and citizens capable of contributing concretely to the construction of a sustainable future. Only through global partnerships that bring together educators, researchers, policymakers, and local communities can we develop a renewed and sustainability-oriented geoscience education.

This collaborative approach will enable us to cultivate a new generation of Earth experts with the technical skills and social awareness necessary to address the complex environmental, social, and economic challenges of our time, transforming geosciences from an often-marginal discipline into a central element of our collective journey towards a more sustainable and equitable future.

Fold in Limestone, near Caminito del Rey, Malaga

Further reading

Gill, J.C. (2017). Geology and the Sustainable Development Goals. *Episodes*, 40(1), 70-76. https://doi.org/10.18814/epiiugs/2017/v40i1/017010 *A seminal article that systematically explores the links between geosciences and all 17 SDGs*

Gill, J.C., & Smith, M. (Eds.). (2021). *Geosciences and the Sustainable Development Goals*. Springer Nature. https://doi.org/10.1007/978-3-030-38815-7 *A comprehensive volume dedicated to analysing the interconnections between geosciences and SDGs, with specific chapters for each goal*

Stewart, I.S., & Gill, J.C. (2017). Social geology—integrating sustainability concepts into Earth sciences. *Proceedings of the Geologists' Association*, 128(2), 165-172. https://doi.org/10.1016/j.pgeola.2017.01.002

Lubchenco, J., Barner, A.K., Cerny-Chipman, E.B., & Reimer, J.N. (2015). Sustainability rooted in science. *Nature Geoscience*, 8(10), 741-745. https://doi.org/10.1038/ngeo2552 *Analyses the fundamental role of Earth sciences in pursuing the SDGs*

Schlosser, P., & Pfirman, S. (2012). Earth science for sustainability. *Nature Geoscience*, 5(9), 587-588. https://doi.org/10.1038/ngeo1567 *Article highlighting how geosciences are essential for addressing sustainability challenges*

Di Capua, G., Peppoloni, S., & Bobrowsky, P.T. (2017). The Cape Town Statement on Geoethics. *Annals of Geophysics*, 60, Fast Track 7. https://doi.org/10.4401/ag-7553 *Document exploring the ethical dimension of geosciences in the context of sustainable development*

Bohle, M. (Ed.). (2019). Exploring Geoethics: Ethical Implications, Societal Contexts, and Professional Obligations of the Geosciences. Palgrave Pivot. https://doi.org/10.1007/978-3-030-12010-8

Mora, G. (2013). The need for geologists in sustainable development. *GSA Today*, 23(12), 36-37. https://doi.org/10.1130/GSATG185GW.1

Senger, Kim2024 Sustainable Development Goals and the Geosciences: A Review 10.3389/esss.2024.10124Earth Science, Systems and Society

CHAPTER 18.
INTERDISCIPLINARY CONNECTIONS:
HISTORY, SCIENCE, AND ART IN THE GEOSCIENCES

Geosciences are everywhere. This simple truth has run through this entire book like a connecting thread. Geosciences are not just a scientific discipline limited to laboratories or field expeditions. Earth – with its rocks, its minerals, its landscapes created by ancient forces – is not only the physical foundation of our life, but also the inspiration for infinite cultural expressions.

This exchange between geosciences and other fields of knowledge creates bonds that enrich both. Literature has made complex geological concepts comprehensible; art has shown invisible processes; video games have transformed stratigraphic principles into interactive experiences. In return, poetic metaphors have helped geologists see phenomena in new ways and artistic insights have revealed aspects of landscape that scientific analysis had not grasped.

This chapter presents examples of this two-way relationship: geosciences have influenced cultural imagination and have been enriched by literature, art, and other expressive forms. These cases demonstrate that the most interesting discoveries arise from encounters between different disciplines, where science and art unite.

Concluding this journey by exploring these cultural connections means showing that geosciences are not isolated but connected to many other forms of knowledge that define our relationship with Earth. It is an invitation to recognise that geosciences, beyond their scientific dimension, possess a profound humanistic value that deserves to be explored and celebrated.

Throughout human history, every culture has developed its own ways of understanding and representing the geological world. While this chapter includes examples from European and North American traditions that have been influential in modern geoscience education, it is essential to recognise that geological knowledge and artistic expression exist in rich diversity across all continents and cultures. Indigenous knowledge systems, Asian philosophical traditions, African oral histories, and Latin American literary movements all offer profound insights into the relationship between humanity and Earth's geological processes.

18.1 Visual Arts and Geological Observation: From Renaissance to Contemporary

Leonardo da Vinci and Raphael were not only great painters but also keen observers of nature. Leonardo's famous “Virgin of the Rocks” does not simply show rocks as a decorative background: the limestone formations he painted are geologically accurate and show how caves are formed through water erosion. Leonardo had studied the caves of Lombardy firsthand and understood, centuries before geologists, how water carves rocks to create underground cavities.

Leonardo was fascinated by the role of water in shaping the landscape. In his notes, he describes how rivers erode mountains and how sediments are formed, theories that official geology would only develop in the 18th century. He had also observed marine fossils in the mountains, intuiting that those lands were once covered by the sea. The grotto of the Virgin of the Rocks thus becomes both a sacred place and a painted lesson in geology.

The Virgin of the Rocks - Leonardo da Vinci

Raphael, in the landscapes of his portraits and Madonna, paints the hills around Urbino with scientific precision. The rolling hills of the Marche region seen in his paintings are not invented: they reflect the true geology of the area, made up of clay and sandstone that give the landscape its soft, undulating shapes. When geologists today study the “Madonna del Granduca” or the “Fornarina,” they recognise the real geological structure of the Marche region in the painted hills.

Both artists observed the territory with a scientific eye, anticipating the methods of modern geology. Their paintings have become valuable documents that today help researchers understand what the landscapes were like in the Renaissance and how they have changed over time.

Geological representation in non-Western traditions

Long before European scientific illustration, other cultures developed sophisticated ways of representing geological features. Chinese landscape painting (shanshui) from the Tang Dynasty onwards captured the essence of rock formations and mountain structures with remarkable geological accuracy. Artists like Fan Kuan (990-1030) studied limestone karst formations directly, and his paintings show appreciation of erosion patterns and stratigraphic layers. Similarly, Japanese artists of the Edo period, particularly Hokusai in his "36 Views of Mount Fuji," documented volcanic landscapes with both artistic beauty and geological precision, recording the mountain's form before and after eruptions.

18.2 Jules Verne and underground cartography

Jules Verne's "Journey to the Centre of the Earth" (1864) is an iconic example of the mutual influence between literature and geological science. Verne did not merely use the geological knowledge of his time as a narrative backdrop but meticulously integrated it by developing a veritable underground cartography that influenced both the collective imagination and speleological research.

Verne's representation of geological strata as temporal archives recording the history of the Earth anticipated the modern approach to stratigraphy. The protagonists' journey through increasingly ancient strata became an effective narrative metaphor for the concept of deep geological time, making accessible to the public a scientific idea that at the time remained abstract even for many scholars.

Verne's influence on speleology is well documented: several expeditions in the second half of the 19th century explicitly cited the novel as a source of inspiration. The French cartographer and speleologist Édouard-Alfred Martel, considered the founder of modern speleology, acknowledged the impact that Journey to the Centre of the Earth had on his decision to devote himself to underground exploration.

Voyage au Centre de la Terre Édouard Riou 1864 Wikimedia Commons

18.3 The reciprocity between fiction and geology: reinventing lost worlds

Arthur Conan Doyle, in The Lost World (1912), created a work that not only popularized Palaeontology but also established a two-way dialogue between science and fiction. Professor Challenger's isolated plateau, where prehistoric creatures survive, was inspired by the real paleontological discoveries of his time. This relationship between science and fiction was not one-way: The Lost World in turn inspired generations of palaeontologists, such as Roy Chapman Andrews, whose expeditions to the Gobi Desert led to the discovery of the first dinosaur egg nests.

The concept of the "lost world" provided a conceptual framework that influenced paleobiogeography, becoming an explanatory model adopted in scientific discourse. Locations such as Lake Tele in Congo and the Pantepui plateau in Venezuela have been scientifically explored following this paradigm.

Original illustration from "The Lost World" by Harry Rountree (1912) Wikimedia Commons

18.4 Michael Crichton and "Jurassic Park": the mutual reinvention of science and fiction

"Jurassic Park" (1990) is a paradigmatic example of how science and fiction can influence each other. Crichton constructed his novel by integrating the most advanced theories of Palaeontology and molecular biology of the time, consulting scientists such as John Horner.

The reciprocal relationship became apparent when the novel and Spielberg's film profoundly changed the direction of paleontological research. After the book's release, there was a significant increase in funding for research on ancient DNA and amber taphonomy. The depiction of dinosaurs influenced Palaeontology: when Utahraptor, a dromaeosaurid similar in size to the cinematic velociraptor, was discovered in 1993, palaeontologist James Kirkland declared, "It almost seemed like Crichton had inside information about the discovery."

18.5 English Romantic poetry: when geology shapes literary metaphors

The English Lake District, with its dramatic geology, served as the true muse of William Wordsworth and Samuel Taylor Coleridge's poetry. In his poem "The Prelude," Wordsworth developed a veritable "geological aesthetic," with descriptions of the rock formations of Borrowdale that reflected a surprisingly accurate understanding of volcanic processes.

J.M.W. Turner - Lake District showing the geological formations celebrated by Wordsworth - Wikimedia Commons

At the same time, Charles Lyell, the father of modern geology, explicitly cited Wordsworth in his "Principles of Geology" (1830-33), even adopting some of his poetic metaphors to explain complex geological concepts. The Romantic sensibility for the "geological sublime" helped to culturally legitimize the new discipline of geology.

18.6 Georgia O'Keeffe: when art reveals what geology cannot see

Georgia O'Keeffe's work, with its depiction of New Mexico rock formations, establishes a profound dialogue with geology. Through her "intuitive geology," O'Keeffe isolated and emphasized structural elements of rocks that revealed underlying geological processes. In her depictions of Ghost Ranch, the simplification of forms paradoxically highlighted stratigraphic features that were difficult to observe.

Pedernal- Georgia O'Keeffe

Several field geologists have acknowledged how O'Keeffe's works helped them "see" the formations they were studying differently. Geologist William Muehlberger stated that O'Keeffe's paintings allowed him to identify previously overlooked tectonic features in the Rio Grande Rift formations.

18.7 - Aboriginal Rock Art and Deep Time

Australian Aboriginal rock art represents one of the longest continuous artistic traditions on Earth, spanning over 65,000 years. The ochre paintings in sites like Kakadu National Park and the Kimberley region are not merely decorative but encode geological knowledge passed through generations.

The ochres themselves—hematite reds, limonite yellows, kaolinite whites—demonstrate sophisticated knowledge of mineral properties and locations. Aboriginal artists travelled specific routes to quarry these materials, and the 'songlines' that guided these journeys are essentially oral geological maps, describing rock types, water sources, and landscape features across thousands of kilometres.

Significantly, some Aboriginal stories describe events that geologists now recognise as actual occurrences: the flooding of coastal areas with sea-level rise at the end of the last Ice Age (around 10,000 years ago), volcanic eruptions in Victoria, and even the formation of crater lakes. These oral traditions preserve geological memory across time spans that exceed written historical records, offering a model of how deep geological time can be communicated through narrative and art.

18.8 - Stone, Jade, and Cultural Geology in Asian Traditions

"The appreciation of stones as objects of contemplation has deep roots in Asian cultures, in particular in China and Japan, where geological aesthetics developed into refined art forms.

Suiseki (Japan) and Gongshi (China): These traditions involve collecting and displaying naturally formed stones that suggest landscapes, mountains, or abstract forms. Collectors prize stones showing evidence of erosion, weathering, and structural features—essentially celebrating geological processes as art. The practice requires knowledge of rock types, formation processes, and recognition of aesthetic qualities like texture (shi), color (se), and surface patterns (wen).

Jade in Chinese culture: Beyond its aesthetic value, jade (nephrite and jadeite) holds profound cultural significance in China. Ancient jade workers developed empirical knowledge of this metamorphic rock's properties—its toughness, workability, and distinctive colors related to mineral inclusions. The geological sourcing of jade along trade routes from Xinjiang and Burma shaped economic and cultural exchanges for millennia.

These traditions demonstrate that geological materials can be simultaneously objects of scientific interest, artistic appreciation, and spiritual significance—a holistic approach that modern Western geoscience has only recently begun to reconsider."

18.9 Pompeian mosaics: geological documentation and cultural memory

The Roman mosaics of Pompeii are an extraordinary example of artistic documentation of geological phenomena. The famous mosaic in the House of the Faun depicting Vesuvius before the eruption of 79 AD has been studied by modern volcanologists as a record of the volcano's pre-eruptive morphology.

The reciprocity is evident in the way contemporary volcanologists use these mosaics to reconstruct the eruptive history of Vesuvius, while new geological studies allow for a correct interpretation of the details depicted. The idyllic scenes of agricultural life on the slopes of the volcano document a period of quiescence that led to a cultural underestimation of risk, a recurring pattern in the history of populations living near volcanoes.

18.10 Robert Smithson's Land Art: material dialogue with geological processes

Robert Smithson's Spiral Jetty (1970) is perhaps the most accomplished example of a two-way dialogue between art and geology. This monumental spiral of rocks and mud in Utah's Great Salt Lake is not only a work of art but also a real experiment in geological processes of sedimentation, erosion, and crystallization.

The spiral, which submerges and reemerges cyclically with changes in the lake's water level, functions as a device for observing hydrogeological changes. Geologists and limnologists have used Spiral Jetty as an indicator of fluctuations in the Great Salt Lake.

The work has also influenced geoconservation. The concept of a 'cultural geosite'—a geological site that is of scientific interest but also has significant historical, artistic, or landscape value—has influenced geological protection policies, recognising that the value of a geological site lies not only in its physical characteristics but also in its cultural significance.

Spiral Jetty- Roberti Smithson

18.11 - Latin American Literature and Geological Imagination

Latin American writers have developed unique literary approaches to geological landscapes, often blending realism with elements that transcend purely scientific description.

Pablo Neruda's 'Alturas de Macchu Picchu' (1945): This extended poem treats the Incan site not just as archaeological ruins but as living geology. Neruda's language captures the granite's endurance, the erosion of time, and the stones as repositories of human memory. The poem exemplifies how geological deep time can intersect with cultural memory and social justice themes.

Gabriel García Márquez and geological magical realism: In 'One Hundred Years of Solitude,' García Márquez describes geological phenomena—rain that lasts years, shifting landscapes—in ways that blur boundaries between natural processes and narrative imagination. This approach, characteristic of magical realism, offers a different epistemology for understanding geological change, particularly relevant in tropical regions where dramatic weather events and landscape transformations occur.

Indigenous cosmologies: Many Indigenous American traditions encode geological knowledge in origin stories. The Popol Vuh of the K'iche' Maya describes mountain formation and the properties of different earths used to create humanity, reflecting empirical observation of volcanic landscapes and soil types in the Guatemalan highlands.

18.12 Art Responding to Geological Disasters: Global Perspective

Alberto Burri's Grande Cretto, created on the ruins of Gibellina (Sicily) destroyed by the 1968 Belice earthquake, is a powerful example of how art can dialogue with catastrophic geological events.

Burri transformed the rubble of the city into an immense work of art: a huge white concrete "cretto" covering 8,000 square meters, following the original topography of the city and preserving the layout of its streets. The artificial cracks in the Cretto visually recall the fractures in the clay, creating a parallel between geological and human trauma.

The Cretto of Gibellina- A. Burri

The work has influenced the way seismologists and geologists' study and communicate the cultural impact of earthquakes. Researchers at the National Institute of Geophysics and Volcanology have developed the concept of "cultural seismic memory" by studying how the Cretto has influenced the perception of seismic risk in local communities. After the 2009 earthquake in L'Aquila, architects and urban planners explicitly cited Burri's approach in the debate on reconstruction, emphasizing the importance of preserving the physical memory of a seismic event rather than erasing it completely.

Similar artistic responses to geological disasters exist worldwide. In Japan, the 'Kataribe' storytelling tradition preserves memory of tsunami events through oral narrative, while contemporary artists like Chim↑Pom created works responding to the 2011 Tōhoku earthquake. In Chile, artists have created memorials for earthquake victims that incorporate geological materials from affected sites. In Haiti, post-2010 earthquake art installations address themes of geological vulnerability and resilience. These diverse responses show how art processes geological trauma across cultures.

18.13 Minecraft: interactive geology and playful learning

Minecraft represents a case of how geology can become the core of an interactive experience that simultaneously entertains. World generation relies on algorithms inspired by real geological principles, creating landscapes with mountain chains, valleys, and mineral deposits that follow recognizable patterns.

The United States Geological Survey has developed educational modules based on Minecraft to teach principles of stratigraphy and mineral extraction. The gaming community has spontaneously deepened their geological knowledge, creating forums with sophisticated discussions on erosive processes and mineral distribution.

Modified versions of Minecraft have been adopted by institutions like the British Geological Survey for scientific outreach programs. The "GeoExplorers" project uses a version of the game that accurately reproduces British geology to teach geological concepts in schools, creating a virtuous circuit where entertainment and geological science mutually enrich each other.

Parallel to this, the comic medium is emerging as a powerful tool for geological outreach. Educational webcomics and social accounts like @geo_logica on Instagram use panels to explain tectonic processes, while the "talking rocks" format anthropomorphizes minerals making them protagonists of stories that explain their origin and properties. Comics offer unique advantages: they can compress geological time into narrative sequences and use visual metaphors to make abstract phenomena comprehensible.

Both media transform abstract geological concepts into concrete and accessible experiences, creating a virtuous circuit where entertainment and geological science mutually enrich each other.

Educational initiatives using Minecraft for geoscience extend globally. In South Africa, the 'Cradle of Humankind' project recreated fossil sites in Minecraft for student exploration. In Japan, students used the game to model tsunami impacts on coastal geology. These applications demonstrate how digital tools can be adapted across diverse geological contexts and educational systems.

18.14 The song of crystals: when mineralogy meets music

In the beginning was the Music of the Spheres, the philosophical concept that mathematical relationships express not only quantitative relationships but also qualities that manifest themselves in forms and sounds, all connected in a huge model of proportions. According to this conception, the movements of celestial bodies themselves produce a kind of music, inaudible to the human ear but consisting of harmonic-mathematical concepts.

The Department of Earth and Geo-Environmental Sciences at the University of Bari, as part of the Sonic Structures project, has developed an innovative method for translating the chemical structural characteristics of crystals into music, in order to obtain a musical composition, i.e., a set of ordered sounds that facilitate the mastery of complex concepts in chemistry, mineralogy, materials science, biology, and on the basis of which it is possible to develop an appropriate musical environment for the museum display of minerals.

Music and crystals. Two different worlds united by a magical word: symmetry. In this project, the order of atoms becomes perceptible through sounds, based on a specially designed scientific model that associates crystallographic parameters with musical ones. The result is surprising: crystals of halite, calcite, diamond, ice, and others become ordered sounds. Each crystalline phase is uniquely represented by a musical composition derived from the crystal-chemical properties of the atoms that form its structure.

This encounter between mineralogy and music has been brought to the public in events such as the conference-concert "Il Canto dei Cristalli" (The Song of Crystals), where compositions derived directly from crystal structures are performed.

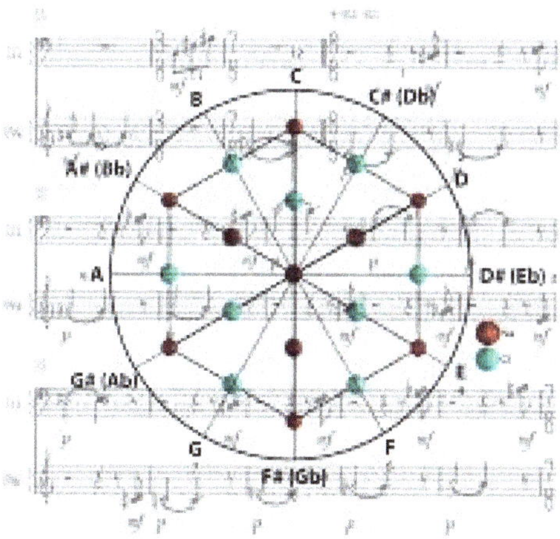

Symmetry relationships between music and crystallographic parameters

The project also extends to other expressions of geosciences: the INGV (Italian-National Institute of Geophysics and Volcanology) has developed "Il Suono della Terra" (The Sound of the Earth), an initiative that translates seismic and volcanological data into musical compositions, allowing us to “listen” to the movements of our planet in artistically meaningful forms.

This sonification of geological and mineralogical structures is an interesting powerful example of how the scientific and artistic dimensions can integrate, creating new ways of understanding and appreciating natural phenomena that transcend traditional disciplinary barriers.

Similar intersections of music and geology exist in other traditions. Indigenous Australian 'song cycles' encode information about water sources and rock formations in musical patterns. In Iceland, composer Jóhann Jóhannsson created works inspired by glacial sounds and volcanic rumbles. The Japanese concept of 'ma' (negative space) in music has been related to appreciate geological voids and karst formations. These examples show how sound and Earth sciences intersect across cultures.

18.15 Conclusion: an evolving dialogue

The examples in this chapter, drawn from diverse cultures and time periods, demonstrate that the dialogue between geosciences and human creativity is truly universal. From Aboriginal songlines encoding 65,000 years of landscape knowledge to contemporary digital explorations in Minecraft, from Chinese jade appreciation to Italian earthquake memorials, humanity has always sought to understand, represent, and respond to geological phenomena.

Importantly, this chapter represents only a small fraction of global geo-cultural connections. Every region, every culture has developed its own ways of engaging with geological reality—ways that often combine what Western tradition separates into 'science' and 'art,' 'knowledge' and 'spirituality,' 'observation' and 'imagination.'

For educators, these diverse examples offer multiple entry points for students from different cultural backgrounds to connect with geoscience. A student in Australia might engage more deeply through Aboriginal geological knowledge; a student in Japan through traditions of stone appreciation; a student in Latin America through Neruda's poetic geology. The universality of geological processes meets the diversity of human responses, creating rich possibilities for culturally responsive geoscience education.

As geoscience education becomes increasingly global, it must also become increasingly inclusive of the many ways humans have understood and represented Earth. The future of the discipline lies not in imposing a single cultural framework, but in creating space for dialogue between different knowledge systems—recognising that scientific and traditional, Western and non-Western, ancient and contemporary approaches all have valuable contributions to make in understanding our planet.

Assale Salt Lake, Dallol, Ethiopia

Further reading

Ballard, J.G. (2018). *The Drowned World*. Liveright Publishing Corporation. https://wwnorton.com/books/9780871404060

Conan Doyle, A. (1912). *The Lost World*. Hodder & Stoughton. https://www.gutenberg.org/ebooks/139

Crichton, M. (1990). *Jurassic Park*. Knopf. https://www.penguinrandomhouse.com/books/337539/jurassic-park-by-michael-crichton/

Lee, D.W., & Gould, K.S. (2002). "Why leaves turn red." *American Scientist*, 90(6), 524-531. https://www.americanscientist.org/article/why-leaves-turn-red

Verne, J. (1864). *Voyage au centre de la Terre*. Pierre-Jules Hetzel. https://www.gutenberg.org/ebooks/4791

Dean, D. R. (1992). *James Hutton and the History of Geology*. Cornell University Press. https://www.cornellpress.cornell.edu/book/9780801480775/james-hutton-and-the-history-of-geology/

Kaiser, B. A. (2018). Developing the 'Geo-social' View in Geography: Encounters and Engagements Within Rocks Community. Progress in Human Geography, 42(5), 686-702. https://doi.org/10.1177/0309132517730452

Norwood, V., & Monk, J. (Eds.). (2016). *The Desert Is No Lady: Southwestern Landscapes in Women's Writing and Art*. Yale University Press. https://yalebooks.yale.edu/book/9780300063011/desert-no-lady

Frodeman, R. (2003). Geo-Logic: Breaking Ground Between Philosophy and the Earth Sciences. SUNY Press. https://sunypress.edu/Books/G/Geo-Logic2

Praet, I. (2021). *Spitting Stones: Volcanology as a Cultural Science*. HAU: Journal of Ethnographic Theory, 11(1), 27-47. https://doi.org/10.1086/713832

Orel, H., & Herlinger, M. (2020). Sonic Crystallography: The Study of Crystal Symmetries through Sound. Journal of Chemical Education, 97(10), 3664-3670. https://doi.org/10.1021/acs.jchemed.0c00483

Thornes, J. E. (2008). *Cultural climatology and the representation of sky, atmosphere, weather and climate in selected art works of Constable, Monet and Eliasson*. Geoforum, 39(2), 570-580. https://doi.org/10.1016/j.geoforum.2006.10.007

Bogost, I. (2011). *How to Do Things with Videogames*. University of Minnesota Press. https://www.upress.umn.edu/book-division/books/how-to-do-things-with-videogames

Borchia R, Nesci O. (2020) Raffaello e luce sia. Sui fondali ritrovati nelle terre d'Urbino.

Aboriginal and Indigenous Knowledge:

Nunn, P.D., & Reid, N.J. (2016). Aboriginal memories of inundation of the Australian coast dating from more than 7000 years ago. *Australian Geographer*, 47(1), 11-47.

Clarke, P.A. (2009). Australian Aboriginal ethnometastronomy and geomythology. *Archaeoastronomy*, 22, 75-90.

Asian Traditions:

Hay, J. (2010). *Kernels of Energy, Bones of Earth: The Rock in Chinese Art*. China Institute.

Addiss, S. (1989). *The Art of Zen*. Abrams.

Latin American Perspectives:

Neruda, P. (1945). *Alturas de Macchu Picchu*. (Multiple editions/ translations)

Saldaña-Portillo, M.J. (2016). *Indian Given: Racial Geographies across Mexico and the United States*. Duke University Press.

APPENDIX: ADAPTING REGIONAL EXAMPLES TO YOUR LOCAL CONTEXT

A Practical Guide for International Teachers

This book draws extensively on Italian and Alpine geological contexts—not as limitations, but as validated methodological templates. The pedagogical approaches transfer universally; the geological content adapts locally. This appendix shows you how.

The Core Principle

What transfers everywhere is the teaching methodology and investigative structure, the progression from observation to interpretation, the integration of multiple data types, and the development of competencies. What you adapt are the specific geological features and formations, place names and geographical contexts, cultural and historical connections, and field sites and accessibility.

Think of Italian examples as proof that these methods work in real classrooms, not as prescriptions to follow literally.

Three Complete Adaptation Examples

Example 1: From Alpine Valleys to Your Mountains

The original example in Chapter 13 compares Lys and Ayas valleys—two adjacent Alpine valleys with contrasting morphology due to different bedrock geology. The transferable method begins by selecting two adjacent valleys with visible geological differences, then comparing their morphology, vegetation, and hazard patterns. Students discover how bedrock controls landscape through integrating multiple map types, progressing from observation to interpretation to prediction.

For the Appalachians in eastern USA, you might replace this with two parallel valleys carved through folded sedimentary rocks—one following a limestone belt showing a wider valley with karst features, the other cutting through resistant sandstone creating a narrow, V-shaped profile. The pedagogical structure remains identical despite different rocks.

In New Zealand, you could compare a valley carved in schist from the Southern Alps with one in greywacke, examining contrasting weathering patterns, slope stability, and vegetation cover. The investigative approach stays the same.

In the East African Rift context, contrasting a volcanic valley with recent basalt, steep walls, and waterfalls against a sedimentary one with gentler slopes and a wider floor uses the same analysis framework.

What remains constant across all these adaptations is that students analyse morphology, identify controlling factors, predict hazards, and integrate multiple data sources. The reasoning process is identical. What changes are the rock types, specific features, local place names, and accessible field sites.

Example 2: From Italian Marble to Your Local Building Stone

The original example in Chapter 7 traces Carrara marble from its geological formation through historical quarrying, use in Renaissance sculpture, to current sustainability issues. The transferable method connects a local rock to its geological origin, traces its extraction history and cultural significance, links it to regional architecture and art, examines economic and sustainability dimensions, and develops students' ability to read their built environment.

In Portland, Oregon, you might use Columbia River Basalt—formed by massive flood basalts fifteen million years ago, quarried since the 1800s, and visible in downtown buildings. You would connect it to Native American stories and discuss sustainable urban geology using the same investigative structure.

For Edinburgh, Scotland, Craigleith sandstone works perfectly Carboniferous delta deposits that built the Georgian New Town, with exhausted quarries now protected as heritage sites. You would link this to Edinburgh's UNESCO heritage status through the same pedagogical progression.

In Jaipur, India, the pink sandstone that gives the city its name offers rich connections—Vindhyan sandstone extracted through traditional methods, Mughal architectural heritage, and modern conservation challenges all examined through the same analytical framework.

The pedagogical DNA stays the same across all these contexts: geology leads to extraction, which connects to culture, which raises sustainability questions. Only the specific stone and its stories change.

Example 3: From Torgnon Landslide to Your Unstable Slope

The original example in Chapter 9 examines a deep-seated gravitational slope deformation in Valle d'Aosta—a slow-moving, post-glacial feature affecting three square kilometres that integrates traditional knowledge with modern monitoring. The transferable framework moves through systematic observation of morphological indicators, integration of historical documentation and local memory, scientific monitoring data analysis, risk assessment and community engagement, and finally development of mitigation strategies.

Applied to California, this same structure works perfectly for the Portuguese Bend landslide on the Palos Verdes Peninsula—the same slow-moving phenomenon in different geology (sedimentary rather than metamorphic), using identical integration of community knowledge and scientific monitoring with the same investigative structure.

In the Himalayan context, Joshimath subsidence in India offers a compelling adaptation—slow deformation affecting an urban area that combines traditional building knowledge with modern geotechnical data through the same pedagogical approach.

For Rio de Janeiro, hillslope instability in favelas integrates geological factors with informal settlement patterns and community resilience strategies using identical investigative methodology while carrying profound social relevance.

The investigation protocol proves universal: observe, document, analyse, assess, communicate. The geological setting and social context vary; the scientific and pedagogical rigor remains constant.

Quick Adaptation Workflow

Step 1: Identify Your Regional Geological Asset

Every region offers significant rock formations or building stones, active geological processes like erosion, tectonics, or volcanism, landscapes shaped by recognizable forces, historical sites with geological connections, and urban geology visible in local buildings.

Step 2: Match It to a Book Example

Match not by identical geology, but by pedagogical function. If you need an example of “rocks tell stories”, use your local building stone. For a "risk assessment" case, draw on your regional hazard. For a "climate change" investigation, examine your retreating glacier, changing coastline, or shifting ecosystem.

Step 3: Keep the Investigation Structure

Maintain the same types of questions moving from observation through analysis to synthesis, the same data integration approach combining maps, images, and measurements, the same progression from concrete to abstract, and the same connection to broader concepts.

Step 4: Add Local Richness

Enhance your adapted examples with traditional or indigenous knowledge from your region, local historical documents and photographs, community interviews and oral histories, and connections to local culture and identity.

The Practical Reality

You don't need to adapt every example in the book. Choose strategically for maximum impact: one major field investigation site near your school, two or three local building stones students see daily, one regional hazard relevant to your community, and your area's signature geological feature. These provide most of the benefit while keeping your workload manageable.

Using Your National Resources

Every country has geological survey institutions like USGS, BGS, BRGM, GSC, and Geoscience Australia providing free geological maps and databases, hazard assessments and monitoring, educational outreach programs, and expert contacts for classroom visits. See the Additional Resources section for links to institutions worldwide. The data infrastructure exists globally—you're adapting proven teaching methods to locally available content.

Why Local Adaptation Enriches

Your students engage more deeply with a landslide threatening their own valley than one in the Alps. Your regional geology is just as valid scientifically and pedagogically as Italian geology. Your adapted examples will be better for your students than the originals because of immediate relevance.

The Italian examples prove these methods work. Your local examples will prove they work everywhere. *Use whatever geology surrounds you—that's where the best learning happens.*

Collaboration Opportunity

Consider connecting with teachers in other regions who are also adapting these approaches. Share your regional versions. Build a global collection of validated teaching materials using the same pedagogical framework with diverse geological content.

This book provides the methodological foundation. You provide the local geological richness. Together, they create powerful learning.

ADDITIONAL RESOURCES –

GLOBAL GEOLOGICAL DATABASES

Primary Institutional Databases

USGS - United States Geological Survey

Earthquake Hazards Program: https://earthquake.usgs.gov/
Volcano Hazards Program: https://volcanoes.usgs.gov/
Water Resources: https://water.usgs.gov/
Mineral Resources Database: https://mrdata.usgs.gov/

BGS - British Geological Survey

UK Geosites Portal: https://www.bgs.ac.uk/geological-heritage/
GeoIndex Onshore: https://www.bgs.ac.uk/map-viewers/geoindex-onshore/
Rock Classification Scheme: https://www.bgs.ac.uk/rock-classification-scheme/

GSC - Geological Survey of Canada

GeoGratis: https://www.nrcan.gc.ca/earth-sciences/geography/topographic-information/free-data-geogratis/11042
Canadian Mineral Occurrence Database: https://smi-msi.nrcan-rncan.gc.ca/

BRGM - Bureau de Recherches Géologiques et Minières (France)

Info Terre Portal: https://infoterre.brgm.fr/
Geological Maps of France: https://www.brgm.fr/en/content/geological-maps

Geoscience Australia

Australian Geological Data: https://www.ga.gov.au/
Online Maps and Tools: https://maps.ga.gov.au/

ISPRA - Istituto Superiore per la Protezione e la Ricerca Ambientale (Italy)

Geological and Geothematic Mapping: https://www.isprambiente.gov.it/en/databases/soil-and-territory/geological-and-geothematic-mapping
SGI Portal: https://portalesgi.isprambiente.it/

Global Specialized Databases
OneGeology Global geological maps portal: http://portal.onegeology.org/ Access to geological services from 120+ countries Harmonized geological data across borders **EarthChem** Global geochemical database: https://www.earthchem.org/ Petrological and geochemical data worldwide Integration with major research initiatives **GPlates Portal** Global tectonic reconstructions: https://www.gplates.org/ Plate tectonic evolution through time Interactive visualization tools
Global Volcanism Program - Smithsonian Institution Active volcanoes database: https://volcano.si.edu/ Historical eruption chronology Volcanic hazard assessments **Natural Earth** Free vector and raster map data: https://www.naturalearthdata.com/ Global coverage at multiple scales Public domain resources International Educational Resources **Earth Learning Idea** Innovative teaching ideas: https://www.earthlearningidea.com/ Multilingual resources for educators Field-tested activities from global contributors

NASA Worldview

Real-time satellite imagery: https://worldview.earthdata.nasa.gov/
Global geological phenomena monitoring
Historical data archive

UNAVCO Education

GPS and crustal deformation data: https://www.unavco.org/education/
Active tectonics study resources
Professional development for educators

SERC-Carleton College

Teaching Resources Portal: https://serc.carleton.edu/teachearth/
Peer-reviewed teaching activities
Professional development modules

International Institutional Equivalents

Function	International Institutions
Environmental Protection & Research	EPA (USA), Environment Agency (UK), ADEME (France), ISPRA (Italy)
Geophysics & Volcanology	USGS (USA), GNS Science (New Zealand), GFZ (Germany), INGV (Italy)
Geological Survey	USGS (USA), BGS (UK), BRGM (France), GSC (Canada), Geoscience Australia, ISPRA-Geological Survey (Italy)
Seismic Monitoring	USGS (USA), EMSC (Europe), GNS (New Zealand), INGV (Italy)

Mineralogical and Petrographic Databases

Mindat.org

World's largest mineral database: https://www.mindat.org/
Over 5,800+ mineral species documented
Global locality information
Crystal structure data

RRUFF Project

Mineral spectroscopy database: https://rruff.info/
Raman, infrared, and X-ray diffraction data
High-quality mineral photographs

Sketchfab - Geological Collections

3D mineral and rock models: https://sketchfab.com/
University collections (e.g., University of Turin, Smithsonian)
Interactive educational resources

Palaeontology Resources

Paleobiology Database

Global fossil database: https://paleobiodb.org/
Biodiversity and extinction data
Stratigraphic information
Community-contributed records

Morphosource

3D fossil models: https://www.morphosource.org/
High-resolution CT scans
Research and education access

TimeTree

Evolutionary timescales: http://www.timetree.org/
Phylogenetic relationships
Divergence time estimates

Global Mapping Portals

OpenStreetMap

Collaborative global mapping: https://www.openstreetmap.org/
Topographic base layers
Community-maintained data

Google Earth Engine

Satellite imagery analysis: https://earthengine.google.com/
Time-series environmental data
Geospatial analysis tools

Regional Resources (representative examples)

Europe:

European Geological Data Infrastructure (EGDI): http://www.europe-geology.eu/
EuroGeoSurveys: http://www.eurogeosurveys.org/
European Federation of Geologists: https://eurogeologists.eu

Latin America:

Geological Survey of Brazil (CPRM): https://geosgb.sgb.gov.br/geosgb/about_geosgb_en.html
Mexican Geological Survey (SGM): https://www.sgm.gob.mx/

Asia:

Geological Survey of Japan: https://www.gsj.jp/en
Geological survey of India: https://www.india.gov.in/website-geological-survey-india
China Geological Survey: http://en.cgs.gov.cn/

Africa:

Council for Geoscience (South Africa): https://www.geoscience.org.za/
Organisation of African Geological Surveys: https://www.oagsafrica.org

www.ingramcontent.com/pod-product-compliance
Lightning Source LLC
LaVergne TN
LVHW052245200926
848481LV00002B/280
9798899663307